Lewis Richard Packard

Studies in Greeg Thought

Essays

Lewis Richard Packard

Studies in Greeg Thought
Essays

ISBN/EAN: 9783337276430

Printed in Europe, USA, Canada, Australia, Japan

Cover: Foto ©berggeist007 / pixelio.de

More available books at **www.hansebooks.com**

STUDIES IN GREEK THOUGHT

ESSAYS

SELECTED FROM THE PAPERS

OF THE LATE

LEWIS R. PACKARD

HILLHOUSE PROFESSOR OF GREEK IN YALE COLLEGE

BOSTON
PUBLISHED BY GINN & COMPANY
1886

PREFACE.

PROFESSOR LEWIS R. PACKARD died on the 26th of October, 1884, in the forty-ninth year of his age, having just completed his twenty-fifth year of service as instructor in Yale College. He was born Aug. 22d, 1836, graduated in 1856, was appointed tutor in 1859, Assistant Professor of Greek in 1863, Hillhouse Professor of Greek in 1867, and became Senior Professor of Greek after the death of Professor Hadley in 1872. He was President of the American Philological Association in 1881, and Director of the American School of Classical Studies at Athens, 1883–1884.

Mr. Packard prepared for the press but two of the Essays in this volume. Doubtless he would have improved the literary finish of the others if he could have revised them, although he was not accustomed to commit his thoughts in full to paper until they were well matured in his mind. The reader will surely not be disturbed by the lack of a rhetorical peroration for the second Essay.

The Essays on Plato are part of a course of lectures prepared for College classes, of which these two only were fully written out, the rest having been given from careful notes with only now and then a finished and elaborated section.

The Summaries of the *Oedipus at Kolonos* and *Antigone* of Sophokles were written at Athens during Mr. Packard's last winter of feverish weakness and suffering (1883–1884), on small slips of paper which he carried in the pocket of his wrapper. One of his few drives during this last visit to Greece was to the hill of Kolonos, and he toiled up the little slope to gaze with charmed eyes upon the beautiful landscape of which he speaks in the Summary.

The jottings which are appended to these Summaries were probably the germs, as they lay in his mind, of such discussions as are found in the Essay on the *Oedipus Rex*.

TABLE OF CONTENTS.

		PAGE.
I.	Religion and Morality of the Greeks	1
II.	Plato's Arguments in the *Phaedo* for the Immortality of the Soul	41
III.	On Plato's System of Education in the *Republic*	65
IV.	The *Oedipus Rex* of Sophokles	77
V.	The *Oedipus at Kolonos* of Sophokles	121
VI.	The *Antigone* of Sophokles	143
VII.	The Beginning of a Written Literature among the Greeks	157

I.

MORALITY AND RELIGION OF THE GREEKS.[1]

I WISH to present to you some thoughts, in the way of suggestion rather than as conclusions, on the morality and religion of the Greeks. It is a topic that has been often touched upon, and in some of its parts treated at great length. I am not so bold as to expect to clear away, at a blow, the difficulties of such a subject, or to advance wholly new views upon it. But it is one upon which new light is continually being thrown, in one part or another, and I may hope that the thoughts which have interested me may interest others also.

It is natural to try to begin at the beginning and see whether we can ascertain what was the basis of the moral ideas of the Greeks. Can we find any pre-existing institution, any simpler or more fundamental series of conceptions, upon which their theories of human duty and their practical rules were founded?

[1] President's address at the annual meeting of the American Philological Association, at Cleveland, July 12, 1881. It was privately printed, and dedicated "to Theodore D. Woolsey, D.D., LL.D., lately president of Yale College, on the fiftieth anniversary of his entering upon the office of Professor of Greek, with most sincere respect and affection, from an old pupil."

It seems plain at the outset that they were not based upon the Olympian theology as set forth by the earlier poets. For that theology during the period of our knowledge of the Greeks was rather out of harmony with the moral sense of the people, lagging behind, as it were, and needing to be corrected and interpreted by the more reflective minds. Thus it has been noticed that the men in Homer are of purer morals than the gods; and it is well known that from Xenophanes on to Plato, and even farther, men are continually criticizing the Olympian theology on moral grounds. And new developments are made of it, reforms within the system, apparently to meet the higher demands of later times. We can hardly admit, then, although it seems to have been a common opinion among the Greeks themselves,[1] that the Olympian theology was the sole or chief source of Greek morality. There must have been some other agency acting alongside of it, to elevate if not to originate moral ideas. Nor could these ideas have been originated by the ceremonial worship connected with that theology, for that is probably itself an effect rather than a cause, and has almost no reference to the larger part of morals, — the duties of man to his fellow-man. The same thing is to be said of the mysteries, if anything can be confidently said of them, and of the oracles with a partial exception, mainly in regard to that of Delphi, to be referred to farther on. I need not linger to prove that the moral ideas of

[1] Isokrates, XI. 41.

the people cannot have been based on the teachings of philosophers. Their task is to explain and defend, and enforce duties already admitted in theory. They are often reformers in morals, but they certainly were not the authors of morality among the Greeks.

Where, then, shall we look for an answer to our question? Was there any other form of belief or practice current among the Greeks which may have contained the germ of moral ideas? There was one, of which the fullest exposition is given by a French scholar, Coulanges. In his work, "The Ancient City," he maintains that the earliest Aryan religion was a worship of the dead, — each family recognizing its departed ancestors as divine beings, and offering worship to them, — and that with this was combined the worship of the hearth-fire, as if its flame was in some sense a representative of the deceased persons. This double worship, he claims, extended through the Indian, Greek, and Italian branches of the Aryan family, lasted throughout the ancient history of Greece and Italy, and still exists in India. He finds his proof in the classical literatures in the shape of references to forms of burial, anniversary rites at graves, and the worship of Hestia. I observe that Sellar in his book on Vergil accepts this theory as well founded,[1] and it must be said that many passages in Greek literature indicate the existence of some such ideas, forming a sort of private family religion by the side of the Olympian system. This worship,

[1] Sellar's Virgil, p. 365 f.

Coulanges holds, was the bond which constituted and preserved the family, and out of the family relation came all the fundamental morality of the people. Duties of kindness and mutual help grew out of regard for the spirits of the dead, truth and purity out of respect for the ever-present deity of the fire. In this last step we cannot follow him, mainly for the reason that before the members of a family could have united in the worship of a deceased ancestor, the family life must have been otherwise developed and been recognized as a bond of mutual rights and duties. If we admit that man has been produced by gradual steps of elevation from animal life, it seems clear that many such steps must have been taken before the custom of ancestor-worship could be established, and that in those steps much of what the theory ascribes to that worship would be already involved. The recognition of descent in a single line and of kinship between collateral branches implies a degree of intellectual and moral development which would leave comparatively little to be done in that direction by the observance of the worship of ancestors. Hence we can give to this worship only a subordinate place in the building up of a moral system. Furthermore, it is by no means clear that this institution or custom was a real worship. It is thought by some careful scholars that it was merely an affectionate honoring of the dead, and it is certainly true that the passages in Greek literature do not clearly show anything more than that, unless in apparently exceptional cases.[1] They do, however, seem to indicate a fixed and constant usage of honor to the dead, which may perhaps fairly be supposed to have had some of the influence which Coulanges ascribes to it under the name of a religion. As to the worship of fire the case is different. We have in Hesiod[2] some important indications of the prevalence of a belief in the divinity of the hearth-fire and the duty of purity in its presence, but in the later life of the people this belief seems to have disappeared or changed its form. It is at least doubtful whether for the Greeks it ever had any such influence or any such connection with the worship of the dead as this theory assumes.

If then we do not find the source of Greek morals in either of these religious systems or in the doctrines of philosophers, perhaps we ought to go back to the time before they left their original seat in Asia, and see if anything in the oldest remains of their Indian kinsfolk can give the answer to our question. We find in Barth's sketch of the religions of India, which I am enabled to pronounce trustworthy on the highest authority, a brief account of the morality implied in the earliest Vedic hymns. Humility, sincerity, affection, in man's attitude towards the gods, benevolence to the suffering, truth and justice in dealings with his fellow-man, — such is the outline that Barth gives, and for the evidence of these ideas of duty, for that which shows these

---

[1] Such as Eur. Alk. 995–1005.   [2] Works and Days, 733 f.

things to have been understood to be duties, he points to the conception of the gods contained in the hymns. Have we here at last found what we are seeking? Not yet, for the question is only pushed one step farther back. Whence came such ideas of the gods? We see in the case of the Greek mythology that it is not necessary for men to have such a conception of beings whom they may worship. How was it then that the Aryans of the Vedic period formed in any degree so pure and lofty ideas of the divine character? It may satisfy us to accept this as an ultimate fact which we cannot analyze, and then we should have an answer to our question: The morality of the Greeks was inherited from their Aryan ancestors, and theirs was founded upon their religion. This answer would once have been enough, but we shall surely be told at the present day that we are looking into the matter at a point too far down the current of history to find the origin of anything, that we must go back beyond all literature to the time of the primitive man, and study in the savage life of some Pacific island or African hut-village the true parallel to the beginnings of Greek life. There can be no objection to such a method from any idea that it would be derogatory to the Greek character to suppose it to have passed through such a period. The Greeks themselves, as full of national pride as any people could be, imagined such a prehistoric stage in the life of their ancestors. Aeschylos makes Prometheus[1] describe men as liv-

[1] Aesch. Prom. 447-471.

ing like ants in holes in the earth, destitute of all the elements of civilization, until he taught them to build houses, to mark the seasons, to count, and so forth. Other poets and philosophers recognize a similar period. But if we adopt this course, we lose our special subject in the wider one of the origin of moral ideas in the human race as a whole, upon which Greek usages may throw light, but only as one among many sources of information. And I think it may fairly be said that, though this method may be the right one, it has hardly yet so proved its processes or led to such definite and accepted results as to justify its general adoption. Unless then we are satisfied with tracing the Greek morality back to the ideas implied in the Vedic hymns and accounting for those as based upon the religious system of the same hymns, I do not see but that we must give up our quest and adopt the words of Antigone[1] when she says of the unwritten laws of religion and duty,

οὐ γάρ τι νῦν γε κἀχθές, ἀλλ' ἀεί ποτε
ζῇ ταῦτα, κοὐδεὶς οἶδεν ἐξ ὅτου 'φάνη.

If now we admit that the origin of Greek morality is lost to our knowledge in the remote past, it is natural for us to look at it within the period known to us and see whether it has a history in that time, whether it undergoes changes either by way of improvement or of deterioration. What then are the materials that we have for this investigation? If we arrange our

[1] Soph. Ant. 456 f. ["They are not of to-day nor yesterday | But live forever, nor can man assign | When first they sprang to being."]

materials in the order of their value, we should put in the first place inscriptions, vase paintings, etc., in a word, all monumental records. These would yield but little information, but that little would be valuable in direct ratio to its scantiness. For they are contemporary witnesses and in a sense impersonal, that is, not likely to be affected by the personality of the author in such a way as to impair the value of their testimony as to facts and usages. We should put next to these, institutions and customs incidentally made known to us by statements in literature, such for instance as the Orphans' Court at Athens, or the practice of offering one's slaves to be tortured for proof of a statement in a trial. As a third source of information, and perhaps the most fruitful one, but needing to be used with critical care as to authenticity and historic probability, and of course with constant observation of dates, we have the recorded incidents of private and public life, all actions of states or individuals of which we can determine the moral character. Such stories should be collected not only from histories but from all the literature, including especially Plutarch, with the aim of forming as complete a picture as possible of the life of the average man. This vein has been worked to advantage by Mahaffy in his "Social Life in Greece," but with certain prejudices and an occasional misuse of authorities which detract from the value of the book. Certainly a great service remains to be rendered by any one who will carefully collect such evidence, without preconceived theories, and present it well arranged and digested. In the fourth and last place would come the deliberate expressions of moral and religious feeling by the poets and philosophers. I put these last partly because they are apt to be put first. The usual way of expounding the religion and morality of the Greeks is to cull passages from the poets and philosophic moralists, to classify those on the same topic together, and thus to frame a scheme of morals which is ascribed to the people at large. This is then offset by evidence of the lewdness of the time, taken generally from Aristophanes, and some glaring cases of cruelty, dishonesty, etc., and we are left with the impression that the Greek character was made up of irreconcilable extremes. But these leading writers are not safe guides as to the moral tenets and practice of the common people, for two reasons. (1) They are picked men, men of profound thought and rich imagination. They may be conscious innovators, leaders in the introduction of new ideas. Some of them, Aeschylos, Euripides, Plato, for example, were at variance with the sentiment of their time and keenly critical of the tone of character prevalent among the people. Plato would have regarded it as an insult to be taken as a representative of the ideas of the mass of men of his day. (2) They are seen in their works at their own highest moral pitch. They are writing under the excitement of poetic or speculative inspiration. They may be writing expressly to instruct and elevate the men about them. They may write better than they themselves ever lived, without any deception, being simply lifted up to a higher plane than they often reached. For these reasons the language of these writers needs to be constantly modified by comparison with the picture of real life to be found in historical narrative or anywhere else. Indeed, an incident casually mentioned by Plato, whether real or fictitious, may be of more value for the purpose in hand than a whole dialogue of lofty moral reasoning. Of course we should not exclude the thoughts of poets and philosophers from our collection of material. The expression of the moral sense of a community takes the most varied forms, and the student of it must pay heed to the extremes in both directions; but yet the most valuable information will come from the comparatively scanty manifestations which lie between the extremes. What he wants to learn are the facts of ordinary life, the actions that seemed natural and so attracted no attention, which for that very reason are rarely recorded and hard to find.

Looking at a part of the period in something of the way now indicated, one might justly say that between the Homeric and the Periklean age there was somehow brought about an improvement in morals. Mr. Grote[1] has pointed out indications of this in three notable particulars,— the position of orphans, the way of dealing with homicide, and the treatment of slain enemies in war. In these there is definite and real progress. In some other respects we find perhaps

[1] History of Greece, Am. ed., II. pp. 91 ff.

less positive traces of the same progress. The family was in the Homeric age established and recognized as the framework of human life. Such a conception as that of Nausikaa is by itself sufficient to prove this. Yet at the same time there are some things not quite in keeping with so high an ideal. For instance, the Greek chiefs at Troy openly keep the captive women as paramours. We can hardly imagine the Athenian generals at Potidaea or Samos doing this in such a way. The rights of property were ill-defined, and especially that of inheritance seems to be not yet securely established. The absence of money and of details of business transactions from the Homeric poems leaves us without means of comparison as to any standard of honesty in such matters. But the honor given to the wily and unscrupulous Odysseus seems to indicate a low morality which as soon as commerce fairly began would show itself fully in that sphere. Without thought of trying to defend the Greeks of ancient or modern times from any deserved reproach in this matter, we ought yet to recognize that the system of exchange and banking which was carried on at Athens in historic times, simple as it may seem in comparison with the modern development, implies a great degree of confidence, which in its turn necessarily presupposes a measure of honesty. The cases of breach of contract or other forms of dishonesty, made known to us by the speeches prepared for the resulting trials, must have been the exceptions, or we cannot see how the system could have come into existence or lasted a week. Again, in regard to courage, as shown in war, there seems to be distinct indication of progress. Though the Iliad is a poem of war, and its pages abound in battles, yet it does not give the impression that military courage in any high degree characterized the heroes celebrated in it or the people among whom it was composed. There is hardly a trace in it of such courage as was shown at Thermopylae or at Koroneia,[1] by which a man can stand at his post and wait for certain death on the chance of saving some one else behind him, or march steadily forward step by step in even line till the enemy's spear touches your breast and the deadly crush comes. Such courage marks a moral advance because it arises from two moral causes: first, a sense of duty, more or less distinctly conceived, to the state or some power above the individual; and second, the habit of disciplined action in a body, which only the influence of some such superior power can originate and maintain. Now it is to be observed that all these indications of improvement in morals are matters which show a development of social relations, an increased sense of society as having claims on the individual and doing work for him. In the treatment of orphans and of homicides the moral sense of the people has substituted for the irregular and uncertain action of the individual or the family a system of definite usage to be followed by some representative of the community. In the treat-

---

[1] See Grote's description, History of Greece, Am. ed., IX. p. 314 f.

ment of enemies slain in war, in matters of honesty and courage, in conjugal fidelity, there is a fuller consciousness of society as standing by and looking on with an opinion that must be respected. There is something of this, of course, in the Homeric poems, but in the later period we see its influence to be decidedly stronger in the particulars mentioned. It is part of the general social progress which is seen as well in government, art, and commerce. On the other hand, there was a decline of morals in some other particulars, two of which may be noticed here. The change in the position of woman in the family is a familiar fact. How far it was due to a greater licentiousness and an increase of luxury and extravagance, as K. F. Hermann[1] suggests, and how far to a change in the political importance of woman, as Mahaffy[2] thinks, we may leave unnoticed here. The form of slavery too shows a change in moral tone. In heroic times, slaves are acquired originally by capture in war, and are regarded as part of the family. In later times they become more commonly articles of merchandise and are used less mildly, as mere machines, in mines and factories. On these two classes the progress in civilization somehow presses heavily to their disadvantage. To the fact above noted, that the advance in morals in the historic time is seen in such matters as belong to a more developed influence of society, another fact corre-

[1] Culturgeschichte der Griechen und Römer, I. p. 135.
[2] Social Life in Greece, p. 136 f.

sponds, that we find in Homer the more private and personal virtues, such as generosity, loyalty to friends, the sense of personal honor, apparently in a better condition than in later times. How far this difference is due to the difference in the sources of our knowledge, may be a question. Of the Homeric society we have a picture refined by the poet's touch, ἐπὶ τὸ κάλλιον κεκοσμημένον, to adapt the words of Thukydides.[1] Whereas in our knowledge of the historic period we come nearer to the hard facts of actual occurrence. Certainly the tendency in a work of imagination is to present ideals of individual characters. The poet will naturally make his heroes and heroines attractive according to his standard, indulging himself in his freedom from the restraint of facts. Here we see a reason to regret our hopeless ignorance of the relative date of the Hesiodic poetry. If, as is supposed, it is but little later than that of the Homeric, then we ought perhaps to take the "Works and Days" as supplying the needed prosaic complement to the heroic ideal, and to form our picture of the early Greek life by combining the two. In that case we might more confidently say that the later historic age shows progress in morals.

It is not difficult to perceive some of the proximate causes of this progress. The gnomes of the wise men, the responses of oracles, the elevated utterances of poets learnt by heart in boyhood and often afterwards recalled to mind,—these all con-

[1] I. 21.

tributed to fix a higher standard. The general advance of the people in the arts of life, the wider distribution of wealth, the establishment of something like a system of law, the facilitation of intercourse between different communities, — all these things helped to make society more refined and to guide the actions of individuals in submission to the general good. Events in history, notably the Persian War, did their part by exciting deep feeling and bringing forth shining examples of heroism. But back of all these there must have been some cause or combination of causes which determined that for a time the progress should be upward and not downward. Why were they able to accumulate and distribute wealth? Why did the arts flourish and law prevail? Why did poets and wise men of such character appear? I do not know that any answer I could give would be other than a modification or an imitation of Bagehot's[1] exposition of the difference between progressive and stationary nations. The progressive nations, to state his view briefly, are such as are able to form for themselves in their infancy a framework of institutions strong enough to hold them together and support their first steps, and at the same time are able also to modify those institutions so as to adapt them to the needs of their further growth. That the Greeks possessed this combination of capacities in prehistoric time is sufficiently evident from the effects and even the linger-

[1] Physics and Politics.

ing remains of it in the period of history. Applying an imitation of this theory to a single part of their complex life, we may say that the Greeks as a people were able to build up a system of usages and of principles based thereon, which supported and shaped, without hampering, the character of the individual. Their sense of proportion and moderation, their love of freedom, their clear-headedness, their power of reasoning on abstract principles, — these qualities, it may be, guided them between a rigid caste system, of which there are some faint traces in their life, and a rude barbarian license. This is only saying in other words, that something in the combination of stock and surroundings made possible for them the attainment of a good result. Perhaps no answer would amount to very much more.

How good was the result? Can we in any degree estimate the value of the Greek system of morals in its best state? Can we say what rank it takes among different systems known to us? If we undertake to do that, two cautions must be borne in mind. (1) We must be careful not to think of the Greeks as exactly like ourselves and to be judged by the same standards. It is necessary to make a real effort of imagination to understand the stock of ideas, the framework of conceptions and assumptions, that was in the Greek mind, before we can rightly estimate the actions based upon that state of mind. (2) On the other hand, we must take care not to think that they were wholly different from ourselves. It is not only that they had the qualities which seem to be wellnigh universal and may be called fundamental in human nature, such as selfishness and avarice, or parental affection and conscience. More than that, they reached a point of civilization, that is, the Athenians and a few other states did, in many respects strikingly like that of modern times. In this fact it is involved that their moral condition, their virtues and their besetting vices, were not unlike ours. It has often been noticed how very modern in some things and how remote in others the life of Athens appears to us when we come to know it a little. For one thing they were very much like us in that their theory of morals was considerably better than their practice. Not only from the professed moralists, but from common men, even from the unblushing scamps on the stage of comedy, we have the most edifying sentiments expressed and immediately forgotten when they come to action. Of course the only proper way to compare the moral conditions of different peoples is to put theory by theory and practice by practice and look at each pair separately. To match the theory of one's own country with the practice of another is simply a cheap self-glorifying. In many respects the theory of Greek morals, if we look at its highest reach, was not very different from our own best theory. That truth was recognized as right and falsehood as wrong, we see in the literature abundantly from Homer through Solon, Mimnermos, Herodotos, the dramatists, down to Plato. So family affection, courage, patriotism, temperance, justice, reverence, — all such virtues are praised and the correlative vices condemned. In some respects, however, there is a difference. In the matter of bodily purity the best standard of the Greeks was low. Revenge is an admitted privilege or duty, until we come to Plato, who first gives a hint of a nobler conception. The passive virtues, such as meekness and gentleness, are ignored. Charity in the form of benevolence we know was practiced, yet we hardly find it inculcated as a duty, unless it is to be recognized in the sacredness of the suppliant. If we look at the general principle of Greek morality, as indicated by some of its best exponents, we must admit that it is a somewhat self-regarding system. It is built up on an idea of fitness rather than of right. It has in some respects a curiously unfinished look, lacking high motives and seeming like an experiment, a tentative sketch of what might be worked up into a grand scheme. As to the other question, how in the practice of its moral theories the community of Athens, for instance, would compare with any modern community, I must confess myself unable to venture an answer. It would require more extensive investigation and combination than I have been able yet to undertake. It seems foolish to enter upon any such comparison with the idea that either of the two objects compared is to be praised at the expense of the other. We ought rather by this time to recognize that different peoples in different periods have differing phases of morality, and to be content with ascertaining the points of distinction without trying to exalt or depress either.

Another question suggests itself at this point. What was the relation of the morality of the Greeks to their religion? How far had the sanction of religion any force to strengthen the moral sentiment? These questions are difficult to answer. They would be so in the case of any people in any age. Consider for instance the English people in the time of the great religious and political struggle called the Reformation, or in the age of Queen Anne, when the question of the succession was so closely involved with the disputes of sects and parties in the Church. How difficult it is, with all our sources of information, in these recent and prominent epochs, to form an opinion how far religion exerted an influence on private life. The opinion is often expressed that there was, certainly as late as the time of Demosthenes, a complete separation in the Greek mind between the ideas of religion and of practical morals. Thus Mahaffy[1] speaks of the Theogony of Hesiod as "showing the changing attitude of the Greek religion by which it was ultimately dissociated from ethics and gradually reduced to a mere collection of dogmas and ritual." Gladstone[2] speaks of the "tendency of the Pagan religion to become the chief corrupter of morality, or, to speak perhaps more

---

[1] History of Greek Literature, I. p. 110.
[2] Quoted by Merry on Od. 8 : 267.

accurately, to afford the medium through which the forces of evil and the downward inclination would principally act for the purpose of depraving it." In a different spirit and with more truth Myers[1] in his essay on Aeschylos says, "Among the Hellenes morality grew up separate from religion, and then, as it were, turned to it to demand its aid." Still more justly Abbott,[2] "The religious conceptions of the Greeks became ethical at an early period and continued to be so to the last, ever growing higher and higher as the conception of life and duty became more elevated." These opinions differ widely enough from one another, yet no one of them can be wholly denied or wholly accepted. Here as before the way to reach the safest judgment is to collect and examine the facts so far as there are facts attainable. At present I can only indicate some of the conclusions which I think such an investigation would establish, although this special topic has never, so far as I know, been fully treated. We may see one form of direct influence in the positive power of oaths. To be sure, they were often violated, but we must remember that it is the violations that attract attention and go on record. The additional sanction given by an oath to a promise or assertion was universally recognized, as appears from the disgrace attached to the name of perjurer. Suicide was looked upon as a sin against the gods; for the effort of the philosopher to explain the theory implies the existence of

[1] Hellenica, p. 15. [2] Essay on Sophokles, Hellenica, p. 38.

the opinion. At least Plato's[1] explanations look altogether towards the gods, while Aristotle[2] speaks only of the injury done to the state. The word ὕβρις in its general use, not as a technical term of law but as a description of a quality of character, includes self-confidence, recklessness, defiance of decency and public opinion, as all having the common element of excess and overstepping due bounds. The conduct thus described, though involving no breach of human law, was yet condemned by common opinion and dreaded as rendering one liable to divine displeasure. Many duties, such as those of hospitality, pity for suppliants, family affection, were enforced by appeals to the god whose titles, ξένιος, ἱκετήσιος, ἕρκειος, show his direct relation to human duty. In such matters as these we see, I think, direct and positive influence from religious belief upon conduct. And I have omitted, you will observe, all those classes of actions which are made immoral by the special institution of religion, such as particular forms of sacrilege, and all such as are condemned by civil law, because I desired to mention only cases wherein religion by itself gave sanction to what all men regard as belonging to universal morality. How should we find it if we look at other matters of daily life still within the domain of universal morality? How far were simple truth without an oath, chastity, courage, temperance, and the like inculcated and practiced from religious motives? Here especially

---

[1] Phaedo, 61 D–62 E.   [2] Eth. Nicom. V. 15.

we should seek the evidence of actual incidents and carefully criticised expressions of sentiment. It would probably indicate that the conception of religion as a distinct motive power available as a sanction of moral duty was not yet fully formed and developed in the consciousness of the mass of men. The two ideas, duty and religion, "We must do what is right" and "Let us worship and obey the gods," were both in the Greek mind. They may have come from different sources. They appear to have had different stages and rates of development. But they approached each other, and at the climax of Greek history they met, at least in some such souls as that of Xenophon and probably other followers of Sokrates. But with the mass of men these two ideas perhaps remained always somewhat separate, very much as they are often kept apart in modern times. It does not seem that the gulf between them was particularly wide in the case of the Greeks, so that no modern parallel to it could be found, yet it cannot be denied that there were elements in the history and spirit of their religion which made such a separation easy and legitimate.

After all, what was the character of the Greek religion? On this subject much has been written and many unwarranted statements made. We are told that it was a worship of beauty; that it was a worship of nature; that it was a mixture of local hero-worship and foreign superstition, with reminiscences of Hebrew tradition and anticipations of

Christian doctrine grotesquely intermingled; that it was a simple and enviable flowering out of human nature unhampered by sense of sin or dread of a future; that it was a profound system of truth, concealing under apparently simple stories the greatest mysteries of the visible and invisible worlds.[1] For each of these and other like statements, there is some show of proof, yet they can hardly all be true. That so many differing views may be taken is due in part to the difficulty of ascertaining the truth. It is difficult enough to frame a clear conception and precise description of any religion held by civilized men, but there are reasons why it is especially so in the case of the religion of the Greeks. It had no standards, no creed, no generally accepted head to control and coördinate local varieties. It was nearly always hospitable to the beliefs and rituals of other peoples, and was itself as composite as the stock of the tribes which made up the nation. It inherited a mythology from an unknown past, some features of which it always retained, modifying only the interpretation of them, and others it expanded and enriched to adapt them to the changes in the civilization and moral sense of the people. It embraced without fatal discord the most widely divergent views and dispositions towards the gods, including in one fold the stern devout Puritanism of Aeschylos and the scoffing obscene Puritanism (strange as this description may

[1] See among others, Preller, Petersen, Gladstone, Symonds, Bunsen, Ruskin.

seem) of Aristophanes. Even in the same mind it allowed the reverent adoration of Zeus and the sublime conception of his nature expressed in the first chorus of the *Agamemnon* to coëxist with the representation of him in the *Prometheus* as being in the early part of his reign a cruel, licentious, and shortsighted tyrant. Of such a religion it seems impossible to get at any central and governing principle, to find any doctrine or spirit which runs through all its manifestations and unites them all.

A religion may be studied either historically or comparatively, either by tracing its own growth through successive stages or by comparing it with other religions. It seems clear that in the case of the Greek religion the former method ought to precede the latter and to control all its processes. For this religion was in a remarkable degree a growing and changing one. Wherever we look at two points in its history, between the Iliad and the Odyssey, between Hesiod and Pindar, between the Persian and the Peloponnesian wars, between Plato and Polybios and on to Plutarch, we still see change. What is true of one period is not true, or true only with many qualifications, of another. A comparison which brought into view only one period of the Greek religion would not be very fruitful; one which neglected the succession of different periods would surely lead to erroneous conclusions. As to the individual deities in many cases there is a history which must be traced out before we can understand the worship, the relations of one deity to another, the local connections. In this field much remains to be done on the plan adopted by Ernst Curtius, Kuhn, Roscher, and others. If we look at the Greek religion as a whole historically, we are carried back at once to a time prior to the existence of the separate Greek national character. We find ourselves obliged to go to the Vedic hymns and try to learn from their scanty evidence what the Aryan religion was. In the nature of the case it is impossible by any such records to reach the very beginning, for the earliest period can leave no record behind it, but it is as far as we can go. In the hymns of the Vedas we find a religious system with a mythology already established. For a brief account of it I depend upon the same authority to which I have already referred, that of Barth. In this Vedic system all parts of nature were held to be divine and were objects of worship. But this is true mainly of what is on the earth and in the atmosphere, for the heavenly bodies are comparatively left out of view. There are numerous deities, some personifications of powers or phenomena of earth and air, in which the physical element has almost disappeared in the personal; others, personifications less complete of abstract ideas or of actions. Each of these in turn is addressed as chief, and the same powers and gifts to men are ascribed now to one, now to another. These deities are represented as acting upon the same motives and subject to the same passions with men. The distinction of sex exists among them, but there is as yet no organized government, nor are they distinctly represented in human form, though the constant ascription of human actions to them implies such forms. They are immortal, and are regarded as lofty and holy beings whom the best of men must humbly worship. It is plain at first sight that this system differs in many respects from what we find among the Greeks at the earliest period when they become known to us, yet on the other hand there are points of resemblance which seem to warrant the belief that the two have a common origin. For instance, certain names of deities the two have in common, although perhaps the only clear examples are Varuna and Οὐρανός, Dyaus and Ζεύς. It is remarkable in both cases that the name prominent as that of a deity in one country is quite subordinate in the other. Οὐρανός has no prominence in Greek mythology, nor Dyaus in Vedic. It is supposed that the early settlers of the Greek peninsula brought with them a form of this worship of the powers of nature. What this form was, — how many deities there were, how fully they were personified, by what rites they were worshiped — we do not seem to have any means of knowing. Herodotos (2 : 52) tells us that the Pelasgians had no names for their gods until they borrowed them from the Egyptians. If we combine this guess of his with the fact that in Homer there is but one obscure reference[1] to an image of a deity, we may infer that the ancestors of the Greeks, like the singers of the Vedic

[1] Il. 6: 273.

hymns, had no representations of their gods and even a less elaborate mythology than they. As time went on, the number of deities was increased, in part by real additions, in part by the re-introduction of the same deity under a new name. Thus Dionysos comes in as a wholly new figure apparently, and even as late as the time of the Homeric poems has not won full recognition. And the Greeks, like the Aryans of the Veda, began to personify human feelings and functions, social principles, and even abstract qualities. On the other hand, Ernst Curtius has traced[1] the progress of the worship of a Semitic goddess from point to point along the lines of trade, whom the Greeks came to know and adopted under several different names, as Aphrodite, Hera, Artemis, and perhaps Athene, with different forms of worship. This multiplication of deities was not wholy due to a mysterious impulse in the Greek mind towards polytheism, but in large measure to an early separation into small communities and a subsequent combination into larger aggregates. Each small community, shut in by its surrounding hills, developed its own form of worship, attaching its own epithet to the common name of the god of sky or sea, and perhaps also deifying its local hero. When intercourse began its work, these all obtained a sort of recognition and a place in the great family of gods. Thus the many wives of Zeus are evidences of so many local myths, which the poets perhaps were the first to gather and combine

[1] Preussische Jahrbücher, 1875, p. 1.

into one story. To the influence of the poets, or to the vivid defining imagination of the race, of which they are only choice examples, is due also the anthropomorphizing tendency which is so prominent in the Greek mythology. As has been already hinted, the difference here between the Greeks and other peoples is only one of degree. All peoples anthropomorphize in some measure. The Vedic hymns ascribe human motives and passions and needs to the gods, but, with a lack of logical sequence, they leave the form of the individual comparatively vague and mysterious. The Hebrew Bible does the same, coming a little nearer in some respects to the Greeks. But the Greeks, obeying at once their reason and their lively fancy, went on and pictured to themselves each god in distinct and beautiful human form. Here came in the plastic and pictorial arts with powerful aid as soon as they grew to perfection. Petersen[1] has remarked how in the age of Perikles sculpture reached its height, just before the wave of skepticism came, so as to fix in the minds of the people the forms of the gods and to provide beauty as a suggestion of holiness. But the arts were only secondary and subsequent in this work. Each deity must have been clearly conceived and defined in form and attributes before the painter or sculptor represented him to the eye. When this was done, it was a great help to the slower minds in imagining the person; but we must not think of these arts as original causes of anthropomorphism.

[1] Ersch und Gruber, I. Griech. Mythologie, p. 155.

One such original cause was perhaps the cheerful, society-loving temper of the early Greek, encouraged by the sunny and temperate skies above him. He easily thought of his god as coming near to him in life and pursuits, and was ready to welcome him if he would appear at his own festival, where a part of the victim was always assigned to him. Every festival, and even ordinary meals, had a religious element. All through the better time of the Greek religion there is a tone of simple gladness, a sort of consecration of physical and social happiness, which may have weakened its moral influence in one direction but must have strengthened it in another. Thus conceiving their gods as individually and socially like themselves, they wrought out in imagination a complete parallel above to their life below, a city in the heavens. The book of Genesis tells us that man was made in the image of God. Aristotle[1] supplies the counterpart to this by his observation that the Greeks made their gods in their own image. It would follow naturally from this that as the character of the people developed and improved, as their theory of an ideal society, their conception of possible excellencies of character, even their knowledge of the extent of the world and the complications of its government, advanced, so would their ideas of the gods be correspondingly elevated. Perhaps the most prominent agents in this upward movement, or embodiments of the spirit that caused it, were the Delphic oracle and the tragic poets of Athens. The part taken by the

[1] Pol. 1, 2, p. 1252 b.

oracle in promoting civilization and elevating in many ways the life of the Greeks has been most elaborately set forth by Ernst Curtius, especially in his History of Greece.[1] There was one special and remarkable outgrowth of the Greek religion, apparently connected closely with the oracle, which, I think I may venture to say, demands more study than it has yet received. This is briefly the belief in Apollo, not simply as the revealer of the hidden will of Zeus, but as the agent of purification to the soul. From this seems to have grown up, if not a formulated system of doctrine, yet a strong faith in the power of the god to bring about an atonement, a reconciliation between the sinner and the divine wrath against sin ; a faith which marks the highest point of practical religion reached by the Greeks. It is most strikingly exhibited to us in the two cases of Orestes and Oedipus. These cases show us also how the tragic poets could contribute to the upward movement of the Greek religion. Plato felt obliged by his theory to exclude them from his ideal state, but it would be hard to find two men who would more heartily have sympathized with his aspirations, when translated from the language of his philosophy into that of their poetry, than Aeschylos and Sophokles. It would also be hard to find two who exercised a wider influence to prepare man for his elevated views than they. The Apolline religion apparently grew out of the Dorian worship of the god, but it found a welcome among the Ionians, and

---

[1] Curtius, History of Greece, Bk. II. Ch. IV.

this illustrates how the oracle at Delphi was one of the main causes of whatever national union the Greeks achieved. It must not be supposed, however, that this prominence of Apollo superseded the Olympian theology. It grew out of that system and came to be the most vital part of it, but never ceased to be a part. Apollo himself is always the son of Zeus and, in this his noblest work, the agent of the will of his father. Zeus remains to the end the supreme god of the Greek religion, and often the expressions used in regard to him, if they stood alone, might fairly be regarded as evidence of monotheism. This was the culmination of the Greek religion, and then of course came the decline. But we must not suppose that the decline began at once. The life of ancient Greece often seems to us to come to an end with the death of Demosthenes and Aristotle. The art, the literature, the philosophy, the free political action, — of all these there seems to be almost nothing after 300 B. C., to interest most of us, and so we are apt to think that the religion too sank at once into a degraded condition into which we need not care to follow its history. But it had a tougher life than they, and there are indications that it continued during the following centuries with undiminished pomp of observance and, if costly offerings are any sign, kept still some hold upon the hearts of the worshipers. Here, more than anywhere, the information as to the actual working of moral and religious ideas is yet to be gathered from inscriptions, institutions, and incidents of daily life.

But we may well believe that the religion was all this time losing its vigor, since the fresh flow of poetic inspiration and the hopeful energy of independent political life had ceased to feed and to sustain it. For this Greek religion had been so shaped by the poets in its growth, and was so involved with the functions and legal rights of the state governments, that the decay or crippling of these two supports must affect it seriously. It would be asking too much of a religion with no higher source than it had, to expect it to do much to preserve the national life from decay when external causes of such resistless power were at hand to destroy it.

Now if some such meagre outline of the history of the matter is in general true, it shows clearly that the Greek religion was not a worship of beauty. This idea seems to have for its foundation nothing but a few instances of semidivine honors paid to persons of striking beauty and the fact that as a people the Greeks were remarkably sensitive to the influence and obedient to the laws of the beautiful. But in reality this quality entered no more into their religion than into their literature and their architecture and all their art. It was for them impossible, we may almost say, to cultivate any form of intellectual or spiritual activity without manifesting in it and impressing upon it their delicate and correct feeling of grace and proportion. Neither was the Greek religion a worship of nature. That was an element in it, at one time, probably the chief element or rather the germ out of which it all grew. But during all the time of which we have knowledge, this original character is lost out of sight entirely, surviving only in a few faint traces, and the names which at the first designated powers of nature have come to stand for a totally different order of conceptions. We might as well say that an oak is really an acorn as that the Greek religion is after all a nature-worship. Nor was it, as we are sometimes told, a display of human nature unclothed and unabashed, acting itself out in the joyous innocent unconsciousness of infancy. From the very first, in order to have a *raison d'être*, it must have recognized the helplessness of man, the dread of an offended superior power, the need of an effort to please an unseen being. And all through the literature of Greece is felt the sterner strain that distinguishes the man from the child, — a sense of duty and of responsibility for the discharge of duty — appearing in Homer, rising to its highest expression in Aeschylos, not wholly lost in Aristophanes, translated into the love of the supreme idea by Plato, and formulated with mathematical precision by Aristotle. Yet once more, we do not find in the Greek myths profound truths disguised as fables. It is possible, no doubt, to read such truths into them and to urge in defense of the practice the use of the same simple stories by the tragic poets as the vehicles of their noble thoughts. But that example does not justify the detection in the myths of wonderful correspondencies with facts not known till centuries later and of the most ingenious selection of names and incidents so as to hide deep thoughts from all but the discoverer and reveal them unerringly to him. The myth of Prometheus is perhaps the most frequent victim of such speculations, and shows what tortures a poor myth, stretched on the rack of hypothesis and torn to furnish food for many more than two fierce constructors of theories, may be made to undergo. But all this interpreting is an inversion of the order of time. Unless these myths were imparted by inspiration from some superhuman wisdom, we cannot reasonably suppose them intended to convey so much profound and abstract truth.

I have been led to speak of the myths, whereas I began to speak of the religion. Is it possible wholly to divorce the two? Is it possible to form an opinion of a religion without looking at the conception which it presents of its objects of worship, its gods, and can we look at the gods of Greece without taking into our view the myths? There are three elements — distinguishable in thought but so closely connected that the discussion of one always tends to pass over into that of another — in the relation of any people to the superhuman world as they conceive it. These are the theology or mythology (that is, the description and history of the divine), the morality (that is, the system of duty among men), and the religion in a specific sense (by which I mean the sanction which the belief in the divine gives to morality). I leave out of consideration the worship as foreign to my subject. Now if we try to estimate the result of these three elements in the case of the Greeks, we find it to be somewhat as follows, I think. The motive cause at the bottom of the whole phenomenon is the need of man for an object of worship above him. That is, for us, a primal need, because we cannot tell whence it arises. Some say from fear, some from wonder, some from sense of sin, some from material dependence. Between these or other causes we have at present no means of deciding, and therefore we may be justified in saying that it is, so far as we can tell, primitive and itself a cause. Πάντες δὲ θεῶν χατέουσ' ἄνθρωποι, says the Homeric poet,[1] — words dear to the heart of Philip Melanchthon. This impulse to worship in the minds of the ancestors of the Greeks produced, if we can trust the best evidence we have, a threefold result, — a worship of the powers and forms of nature on earth and in air, a worship of fire, and possibly, a worship of the dead. This inheritance the migrating tribes brought with them to their settlement in Europe, and in course of time it seems to have become localized and humanized and systematized. They expanded it by adopting deities and beliefs and ceremonies from foreign sources. They added also deities whose names seem to indicate a native Greek origin, such as Themis, Peitho, Metis, and other personifications of qualities or processes. They were ready to see the divine agency all about them, or, in other words, they were, with some notable exceptions, in an uncritical

[1] Od. 3: 48. "All men have need of the gods."

state of mind as to the authority of the prevailing belief. Their conceptions of the gods, clear and clean-cut as they were in some respects, were in others vague, elastic, and constantly open to unconscious modification. There was moreover a sense of good-fellowship, so to call it, in much of their intercourse with the gods, which it is hard for us fully to understand and yet necessary to include in our view if it is to be a true one. The gods were thought to sympathize with men and help them in all their experiences of joy or sorrow, in mere sensual pleasure as well as in the highest intellectual or moral activities. But all along from the beginning, or perhaps from some later point and cause now unknown to us, the conception of these divine beings was just sufficiently above the moral standard of the average man to exert some control upon him and to help him and through him the community up to a higher level. We cannot doubt that Aeschylos believed that the Zeus to whom he prayed, whatever he might have been in an earlier period of his reign, was when he prayed to him, a being wiser and better than himself. We cannot doubt that Plato felt that "the Idea of the Good" was continually lifting him up to better thoughts and a nobler life. Yet each of these men formed in his own mind the conception of the being to whom his worship was offered. There is no marvel or self-delusion in this. We know that an idea may come to any man, with or without recognized external suggestion, which may make his life and character ever after purer and better than it was before. In this way we can understand how the religion of the Greeks was elevated by the improvement of the character of the people and how at the same time it was continually helping to elevate in its turn the character of the people.. Certainly the ordinary citizen of Athens did not habitually think of the gods as Plato and Aeschylos did in their loftiest speculations, but he must have thought of them as above himself in some degree, and he must have been helped to higher views by what he could hear and understand of the thoughts of the great minds.

But here there occurs a difficulty. What shall we think of the worship of Dionysos and that of Aphrodite? They seem to rest upon the deification of two degrading sensual passions which can only lead to the indulgence of vice, so that they are in a word the consecration of vice, and how could there be anything elevating in them or in a system which tolerated them? It would be foolish to try to defend these practices or to explain them away by imagining a theory of morals which would justify them. One thing however can and ought to be said to qualify in some measure the impression they make upon us as to the character of the people among whom they prevailed. It appears altogether probable that both these forms of worship were introduced from other countries and that there was originally nothing corresponding to them in the native Greek belief. "But they were adopted by the Greeks." Yes, they were adopted, at first perhaps because a divine element, akin to that of other nature-powers, was recognized in the myths connected with these deities. Then we may suppose that the worship of Dionysos in particular became so prominent and popular in part because it harmonized so well with the festival meetings of neighbors and friends which the Greeks from our earliest knowledge of them were accustomed to connect with religious observances. It has been observed that there are traces and relics of a worship of Aphrodite in which bodily purity,[1] and a worship of Dionysos in which sobriety,[2] was required of the worshipers. The gross abuses which became associated with the worship of these deities were simply the indulgence of low passions under the pretext of religious service. It is folly, as I have already said, to try to construct a theory of the innocent deification of everything in human nature which will hold good for the Greeks at the culmination of their civilization. Indulgence of such passions was not indeed condemned by them in their best estate so strongly as it has been in modern times, that is, within the last hundred years, but yet, if we can trust at all their literature and the evidences of character in their history, we must admit that it was condemned. We cannot too positively believe and affirm that such excesses were not the legitimate product of a distorted idea of religion, but the abuse of a natural and right idea.

[1] Preller, Griech. Mythologie, I. p. 268.
[2] K. F. Hermann, Culturgeschichte, I. p. 68.

The history of the Christian church can show abuses not less gross but only less public and more inconsistent with its general character.

So then, to conclude, if we look thus at the religion of the Greeks, we see in it a natural development, a close connection with the character and history of the people, a steady progress towards a not unworthy ideal. Compared with Christianity in its highest forms, compared even with Buddhism and Mahometanism in some particulars, it appears wavering in its conception of the divine being, feeble in direct moral influence, and much too tolerant of gross vice. Still I believe it was a religion, and not unworthy of the name, that is, it was a system of belief as to the relation of man to the supernatural world, which influenced him in his conduct and influenced him in a continually increasing measure towards reverence, integrity, temperance, justice, and good-will to his fellow-man. It was more social and external in character than agrees with the highest type of religion, but it must have had even to the common man a personal element and the effect of an inward control, or I do not see how we can account in any reasonable way for the existence of the civilized society of Athens and for the character of Sokrates.

II.

PLATO'S ARGUMENTS IN THE *PHAEDO* FOR THE IMMORTALITY OF THE SOUL.

THE thoughts of such a mind as Plato's on such a subject as the immortality of the soul ought naturally, it would seem, to be of interest to all students of the history of mankind. They should not be expected, of course, to be cast in exactly the forms of thinking of our own day, but the differences which we find in them are just such as to make the study of them interesting and stimulating. In many respects they are remarkably modern; in others we may find that though the form is different the substance is the same. It is often the case that truths have to be restated in a new form for each successive generation of thinkers. The same idea presents itself differently at different times, and sometimes even what may appear strange to us when said by one of a former generation is really what we ourselves are thinking and stating in a form of our own. In the following pages the attempt is made to state with the utmost exactness the several arguments that Plato advances to prove what he wishes to believe,—adding a few words of criticism, only so much as seems needful to the understanding of his thought. There has been some difference of opinion as to the number of his arguments. Some, finding the doctrine of Ideas present in several of these arguments, reduce the number to one, or at most, two. Others, mistaking the answer to one of the objections for an independent argument, raise the number to five. We have followed the plain indications in the structure of the dialogue in presenting four arguments.

I. Plato's first argument is a probability based upon the doctrine of procession of contraries each from its opposite. There is an old doctrine, he says, that the living proceed from the dead. As we know that the dead proceed from the living (*i.e.*, the living pass into, become, the dead), then if the old doctrine be true, this is a case of the alternation of opposites, a sort of cycle passing continually from one extreme to the other, and so back and forth. The probability of such a circular movement he proceeds to prove in three ways, — by analogy, by the presumed symmetry of nature, and by a *reductio ad absurdum*.

(*a*) By analogy. Here he gives a number of instances, first of opposite pairs of things, such as good and bad, just and unjust, and then of evident and necessary passage from one of these to the other. Thus a thing cannot become larger except by passing out of a previous state of being smaller, nor heavier except by passing out of a previous state of being lighter, *etc.* Here we may remark the advantage to his argument from the use of comparative adjectives; but the real strength of it is in the use of the word *become.* It is of course true that nothing capable of growth can become large or small except from being comparatively small or large previously. The fact that he confines his statement to attributives is justified by his regarding living and dead as attributes of man.

(*b*) By the presumed symmetry of nature. Here he confines himself to the case in hand, bringing in no illustrations. We see that the state of being alive exists, and the state of being dead; and equally evident to us is the transition in one direction, from life to death, that is, the act of dying. Therefore if nature is not to be unsymmetrical (*lame* is his word), must we not assume the existence of the opposite transition, from death to life, that is, the act of coming back to life? This is plainly necessary in order to complete the supposed cycle.

(*c*) By a *reductio ad absurdum.* Suppose that there were no such return to life; or, to take first an illustration, suppose that men fell asleep but there were no waking up again; plainly all men would in time be asleep. So if there should be, as we see there is, the transition of dying, but not the corresponding transition of coming back to life, everything would ultimately be dead. But what is the absurdity in this? There is none apparent, unless we supply the thought which, though not expressed, seems to be assumed in the writer's mind, *viz.*, that it is contrary to reason to suppose that this world of life and activity has come into being with no other end in prospect for it but to fall asleep and sink away into universal death.

Thus, then, the ancient traditional belief that the souls of men pass from the world of the living to the world of the dead, and return again thence to the world of the living, is accepted. It is easy to see how such a belief may have arisen; the alternations of life and death in the vegetable world during the progress of the seasons, would naturally suggest it. It was apparently involved in the Pythagorean and Oriental doctrine of transmigration of souls. It would find support, too, in the Herakleitean doctrine that *being* is really a continual *becoming*, that all things are in a perpetual flux. But in the form in which Plato presents it, it rests in part upon another Pythagorean doctrine, namely, that of a sort of polarity in the universe, whereby all things may be grouped in two classes of opposites.

Two or three remarks may be made upon this argument before we speak of its general value. (1) This argument implies a limited, unchanging number of souls in existence. The possibility of the creation of new souls or their production out of any existing materials would destroy the whole argument. Indeed, Plato recognizes this possibility, in order to exclude it, when he says (72 D), " For if living things proceeded from the rest of the universe (*i.e.*, not from the dead), and then should die, the result would be that everything would end in death." The belief in a limited number of souls is distinctly stated by him in the *Republic* (611 A). (2) This argument implies the pre-existence of souls, a belief which appears more prominently in the next argument, and is elsewhere avowed by him. (3) It does not appear from anything in this argument why it should not apply as well to the animal and plant as to the man, and perhaps (*cf.* 70 D) Plato would not have objected to such an extension, as Aristotle would not.

Of the value of this argument, in the view of modern thought, it is not necessary to say much. We do not really know much more about the origin of the human soul than Plato did; but few men, unless a Thomas Beecher, would soberly argue that the souls of the dead return again to this life in the form of bees or wolves, to use Plato's illustrations, or of men. But it may be observed that the argument regards the soul as an independent something, of which being in this life and being in the state beyond death are merely two conditions. It depends wholly upon the assumption that being alive and being dead are two exact opposites, and that there is no third possibility for the soul. If there is such a third possibility,—for instance, destruction,—the argument breaks down. This weak point is seen further on in the discussion, and an attempt is made to cover it.

II. The second argument has much more real substance than the first. It is sometimes concisely described as the argument from reminiscence, which serves well enough for a title; but it should not be supposed that the argument is simply this, that we seem to recall things from a former state of existence, and therefore must have passed through such a state.

That it really contains more than this, will appear from the following formal statement of it.

1. Learning is really recalling or reminiscence. (This is shown in 73 A, by the usual proof that a skilful questioner can lead another person to state a number of truths about a subject on which the person questioned previously supposed himself to have no knowledge.)

2. Reminiscence implies three things, *viz.* : —

(*a*) Previous knowledge of the thing recalled ;

(*b*) That one thing may recall to the mind another though entirely different thing ; *e.g.*, the sight of a lyre may recall the image of its owner, *etc.* ;

(*c*) When the object suggesting and the object suggested are alike, judgment as to the degree of likeness.

3. There exist certain absolute essences, such as beauty, goodness, equality, *etc.*

4. We get our knowledge of them from the senses, *e.g.*, the Idea (= absolute essence) of likeness (τὸ ἴσον) is suggested by the sight of things like to one another.

5. But we also observe at the same time that the Idea of likeness is not perfectly realized in any two like objects, *i.e.*, no two like objects are precisely alike.

6. In order to be able to pass this judgment, that the Idea of likeness is not perfectly realized in any object of the senses, we must have had the Idea of likeness in our minds before we could make the comparison between it and like objects.

7. On the other hand, the perception of the objects compared with the Idea is gained only by the use of the senses; and at the first use of the senses we are in possession of the Idea which we compare with them.

8. But the use of the senses begins at birth; hence before birth we must have had the Idea of likeness. (All this applies to all the Ideas or absolute essences, that of likeness being merely taken as an example.)

9. Hence, as to all such Ideas, there are two alternatives:—

(*a*) Either we are in possession of them from birth all along through life;

(*b*) Or we have lost and are obliged to recall them.

10. We are not all of us in possession of them, for we cannot all of us explain them, and no man knows what he cannot explain.

11. Hence it must be that we have lost them; and the process of learning about them is a recalling of what we knew before in a state before birth.

12. Therefore our souls existed and had intelligence to apprehend these Ideas before birth.

This doctrine of reminiscence is prominent in Plato's system. It does not appear at all in Xenophon, and hence was probably unknown to Sokrates. But it is inseparably connected with Plato's doctrine of Ideas. For the Ideas, being the one unchanging real existence, are the link not only between the seen and the unseen, but also between past and present (and here in the *Phaedo* recognized also as the link between present and future). No one of the ever-perishing objects of this visible world has enough reality of existence to have belonged to that past state; it is only the Ideas that exist in and by themselves, and hence are always to be apprehended in any sphere by whatever being is present capable of apprehending them. The doctrine of Ideas was to Plato the way into the realm of truth, the only attainable theory of knowledge. It was his means of getting a foothold of ground for thought to start from. The Eleatics had denied all existence except that of absolute Being, a pure abstraction having no possible connection with the world of sense. Herakleitos had denied any existence except the process of coming into being and passing out of it, the unceasing flux which every object of sense undergoes continually. Neither of these theories is a satisfactory basis for knowledge or for reasoning. To gain that basis, as well as to overthrow the dogma of Protagoras, *omnium homo mensura*, Plato conceived the perfect original of all sensible objects as existing in a world remote from this, beyond all sense-knowledge, of which world alone real existence could be predicated. The connection between these real existences and the phenomenal existences about us, necessary in order to be able to determine the relation of our world to existence, was not established by Plato very satisfactorily. He was in doubt about the nature of it, and leaves it uncertain (*Phaedo* 100 D) whether it should be called a presence of the Idea in the sensible object or a communion of one with the other or by some other name. But however joined to sensible objects, beauty and goodness and equality and trees and tables, yes, even dust and filth, have real existence only beyond the sphere of the visible in a realm apart. This was the world of Ideas, and of that world, as beyond the reach of sense, the mind could have knowledge only by having been in it during a previous life.

In general it is true enough to say that these Ideas of Plato's correspond to what we call abstract ideas, though he did not use that adjective because it implies a theory as to their origin which he had not formed. We say that such conceptions as those of whiteness, beauty, goodness, are formed by abstracting the common element found in all white or beautiful or good objects from the qualities peculiar to the individual or the sub-class. This may account for the existence of abstract ideas, but it leaves unaccounted for the existence in our minds of the faculty of abstraction. If then we seek to translate Plato's thought into modern thought, we must take a different class of conceptions from abstract ideas. The true parallel in this respect to his Ideas would be such things as the idea of cause, the idea of duty, the conception of space, and others like these, which we find existing in the human mind at its earliest activity and do not know how to account for. With this substitution we should state his argument thus: "We find in our minds certain things for the presence of which we cannot account; the objects of sense around us furnish the occasion of our conscious use of these things, but cannot have originated them; at no time since birth have we been put into possession of them; hence we are born with them; they must be ascribed to some extra-mundane source, and we must suppose that we acquired them in a previous state of existence." All the steps of this argument, except the last four, would be valid in the view of all modern thought but that which denies independent existence and divine origin to the human soul. These last steps were the only conclusion to the argument which could be expected from even such a mind as Plato's in the age in which he lived.

As an argument in the series, it will be noticed at once (as it is in the dialogue), that it may prove prior existence for the soul, but does not at all go to show anything more than the possibility of existence after death. This defect has to be made good by subsequent proof. As compared with the first argument, it has not only the advantage of being based more directly on the facts of human nature, and so of having more logical substance, but also it makes an advance in that it applies only to man, and not to animals or plants as well; and only to the spiritual part of man, and not to his body also. That is because it rests entirely upon intellectual phenomena, and the fact is recognized by Plato in the significant addition (76 C), " Our souls then existed before they came into human form, *apart from the body, and had intelligence*" (χωρὶς σωμάτων, καὶ εἶχον φρόνησιν). At the end of this argument (76 D E), Plato states the substance of it in a pregnant form which is worth repeating, "The existence of our souls before we came into this life is as sure as the existence of the Ideas." Jowett, in his Introduction to the *Phaedo*, points out that this is as if we should say, "The immortality of the soul is as sure as the existence of God," or, "I believe in the existence of God, and therefore in the immortality of the soul."

III. The third argument moves in a sphere more familiar to the popular forms of thought concerning the soul; there is little in it which is not found in ordinary writing of average minds on the subject; cycles of existence, transmigration and a pre-natal life appear no more. We may remark by way of introduction that it is apparently suggested to the mind of Sokrates in the dialogue by the words ξυνίστασθαι ἀμόθεν ποθέν (*compacted from some quarter or other*) in 77 B.

1. An uncompounded thing is probably incapable of dispersion.

2. The always-same thing is probably uncompounded.

3. Ideas are always-same things and invisible, whereas objects about us are ever-changing and visible.

4. Soul is itself invisible, and is always hampered by the body in its efforts to reach by thought the invisible.

5. Hence soul is like and akin to the always-same ideas.

6. Soul dominates body as the divine (and immortal) does the (human and) mortal.

7. Hence soul herein is like the divine.

8. Thus the soul is like the divine, the invisible, the always-same, the indissoluble.

Here the argument might well have stopped, but Plato, perhaps with a view to introducing the next step in the dialogue, goes on to add what rather weakens the foregoing by suggesting mere comparative durability on the part of the soul.

9. The body lasts some time after death, especially if it is embalmed; and some parts of it, such as the bones, seem almost imperishable. Can it be that the higher, purer essence of the soul is less long-lived?

As has been said, this argument moves in a plane of thought familiar in almost every respect to our modern thinking. In the popular conception, death is commonly regarded as a dividing of (1) soul from body and (2) body into its constituents. "Divide and conquer" might be its motto. Hence if it could be proved that the soul, regarded for the moment as in some sense material, is uncompounded, we might infer that it can defy death or any known form of destruction. An atom of matter we suppose cannot in the course of nature be destroyed, and the soul, if simple, would have at least the eternity of matter. Plato seems to mean to distinguish soul from matter by calling it invisible and unchanging, but no such distinction would be admitted now. We may all argue that the soul is invisible and that it has dominion over the body; but when we are further told that it is simple and not compounded, we are inclined to ask whether the predicate is applicable, whether it would not be as proper to speak of a white smell or an oblong thought as of an uncompounded soul. At best, the argument as Plato puts it, does no more than establish a presumption. The likeness of the soul to a class of invisible, unchanging, supreme, immortal objects does not prove that all these predicates are equally applicable to it.

Here the discussion takes a new turn by the introduction of two objections or difficulties suggested by the two Theban friends of Sokrates. The second of these leads to the introduction of Plato's fourth argument, and so will come legitimately into this summary. The first objection, however, is answered by criticisms which contain no new argument, and hence it might be passed over here. But the objection is in itself so akin to the fashionable modern view of the soul, that it seems worth while to give it a little space. The objection of Simmias is suggested by the remark Sokrates had made, that the body as a whole lasts a while, some parts of it a very long time, after death, and that it can hardly be that the pure, invisible essence of the soul does not last longer. Simmias says: "How is it in this parallel case? A strain of melody is invisible and incorporeal and admirable and divine, while the instrument that produces it is earthly and compounded, and akin to the mortal. Shall we say then that, because the broken strings and wood of the lyre last a long time, the melody must be lasting still longer? Now we regard the soul as a melody or harmony, produced by the exact tension of the body in equilibrium between opposing forces. May it not perish when this nice adjustment is broken down, just as the music does, though the materials which were so adjusted last for some time?" The answer of Plato is as follows:—

1. This objection is inconsistent with the doctrine of reminiscence, and with the pre-natal existence of the soul which that doctrine has been shown to imply. For a harmony cannot exist before the material causes of it exist in a state of proper adjustment; hence if the soul is a harmony, it cannot have had an existence prior to this life. But the doctrine of reminiscence is inseparably connected with the doctrine of Ideas, and so must be true.

2. A harmony is determined in its nature and quality by the material things which produce it; hence one harmony is more a harmony than another, if its material causes are better adjusted. But one soul cannot be more a soul than another; hence the soul is not a harmony.

3. One soul may have more of virtue or vice than another. But if those who call the soul a harmony also call virtue harmony and vice discord (which probably, but not certainly, they would do), then, as one

soul cannot be more or less a soul, or, in other words, cannot be more or less harmonized than another, all souls must have an equal degree of harmony in the sense of virtue; or rather, no soul could have any discord, that is, vice. Hence the soul cannot be a harmony.

4. (Recurring to the former principle, under 2, and applying it differently.) Harmony is dependent on the material things that produce it. But the soul leads, opposes, disciplines, chastises the body, as is illustrated by a quotation from the *Odyssey*. Hence the soul is not a harmony.

The objection of Simmias is based upon what seems to have been a current metaphor, probably of Pythagorean origin. Thus it happens that the answer to it put into the mouth of Sokrates is mainly a criticism of the metaphor, and contributes no positive argument for immortality. But the objection itself, though thus easily disposed of, is yet, as has been already said, one of the most modern things in the book. Who does not as he reads it recognize the familiar tone, the view of the nature of the soul which now meets us everywhere, the only view we are told for which there is any evidence at all? There is little difference, from one point of view, between calling the soul a harmony, produced by the action of balanced forces upon the body, and calling it the product of molecular change, or rather, not a product at all, but a mere series of such changes. As Demokritos has been summoned from his sleep of ages to be the patron of one modern theory, so might Herakleitos be brought up from the dead to give authority to this one. For he thought that all existence was but a series of changes, a perpetual flux, and on this theory the mind and soul of man is a mere stream of states of consciousness, like a river passing through the same bed, but never for two seconds together the same actual substance. We cannot certainly say how Plato would have met this theory if it had been formulated in his time, but the points of his criticism on the metaphor of a harmony suggest a probability. The first point above needs but little adaptation in order to become the question, "How can a series of molecular changes have a memory?" And the last point he makes (4) is as good against the modern theory as against the ancient metaphor. It is an appeal to consciousness and to conscience; to consciousness as testifying to the action of the will, to conscience as sitting in judgment upon the decisions of the will.

We come now to the objection of Kebes, which introduces the last and most important of the formal arguments for immortality. Kebes begins by conceding that the soul is much more durable than the body, and therefore may probably enough survive the experience of death. But who can tell, he argues, — but let us have his illustration first, for the sake of its quaintness. He too, like Simmias, will present his thought in a garb of imagery. "Suppose a weaver dies and is buried, and some one brings us his clothes

and says, 'Here are the clothes, still in existence;
must you not admit that the man himself, a much
more durable thing than a garment, is still in existence, too?' Not so, we should answer; for this
weaver made for himself, and wore out, many a garment, and finally perished before the latest garment
was worn out. May it not be, then, that the soul
likewise wears out many bodies, if the man has a long
life, constantly renewing its bodily vesture as it constantly wears it out, but at last finds its own life ex-
hausted in the process, and so dies before the last
form of its body does, leaving it still existing? May
it not even pass through several lives, surviving several
successive deaths, but at last, in some one death (and
no man can tell when that one will come), itself also
perish?"

By way of preliminary to the answer to this objection, Sokrates is represented as giving a sketch of his
own intellectual history, so far as it may be traced in
the effort to determine the true meaning of the idea
of cause. If this sketch could be taken as a real piece
of history, representing what actually occurred in the
case of Sokrates or of Plato, it would be of exceeding
value and interest. But, unfortunately, it does not
seem to be personal history. It begins too far back
for Plato, and goes too far forward for Sokrates. We
must rather look at it as a brief outline, in this bio-
graphical form, of the course of Greek philosophy in
the discussion of the nature of cause. He says that
when he was young he used to be much interested in

questions as to the cause of the origin and of the decline of this thing and that, as, for example, of thought, and of the movements of heavenly bodies, and of the growth of the human body. But from all such inquiries he could get no answer which satisfied him. For instance, if you say one man is taller than another by a head, is it the head that makes him so? If so, then is it the head that makes the second man shorter than the first? But how can the same thing make one man taller and another shorter? Again, division makes two things out of one, and the addition of one to another makes two; but how can division and addition cause the same result? In the midst of these puzzles he heard one day a man reading from a book, said to be by Anaxagoras, that the organizer and cause of all things was Mind. This phrase pleased him wonderfully, suggesting a possibility of all sorts of rational explanations of different phenomena, and he lost no time in getting the book into his own hands. But how wofully he was disappointed when he found that the author of the new theory himself did not make any satisfactory use of it, but went on in the old way, suggesting all sorts of proximate and occasional causes. It was, he says, as if some one should say that he, Sokrates, did everything by (reason of) his mind, and then should go on to say that the reason why he was sitting there in the prison was because his body was made up of bones and sinews, and that, by certain contractions of the latter, he was held there in a sitting position; thus ignoring the fact that his body would long since have been away in Megara or Boeotia, if he had not thought it more just and honorable to stay there where he was. What he ought to have said was that the possession of these bones and sinews was the necessary condition of his sitting there; but Sokrates's judgment of what was best was the real cause. This confusion of necessary condition with real cause was responsible for many of the absurd theories as to natural phenomena prevalent in former times. It seemed necessary, in order to avoid these errors, not to look at realities directly, lest the mental sight should be dazzled (just as the eyes would be if one should look directly at an eclipse of the sun instead of at a reflection of it), but to turn the attention upon conceptions or the world of Ideas, and study there the true reality. The method in this sphere of thought is to determine a principle which is known to be true by its application to a number of cases, and to hold fast by it, and use it consistently; and if it is assailed, strive for some more general principle in its stead, until you reach something which will hold good.

This brings us to the fourth argument.

IV. 1. The existence of Ideas is assumed as a starting-point. (This is the "principle" which is now to be applied to determine the nature of cause.)

2. Objects have qualities by being connected in some way with the Ideas. *E.g.*, It is by partaking of beauty, that an object is beautiful. Or, it is not the addition of one to one that makes two, but the presence of the duad.

3. Opposite Ideas cannot be present in the same object at the same time. It may appear so sometimes. *E.g.*, A man midway in height between two others may seem to have at once smallness and greatness, being smaller than one of the two others, and larger than the other. But these are merely comparative greatness and smallness, not the absolute Ideas. Absolute greatness and smallness cannot co-exist.

4. There are certain objects which contain one Idea in such a way that they cannot admit the Idea opposite to the one they contain, and therefore may always be described by the term which describes the Idea contained in them. *E.g.*, Snow contains the Idea coldness, cannot admit the Idea heat, and so may be always called cold. The number three contains the triad (*i.e.*, the Idea of three), and also the Idea oddness; it can never admit the Idea evenness, and may always be called odd. Each of these, on the approach of the Idea opposed to its contained Idea, must either withdraw or perish.

5. What sort of objects are the foregoing? They are such as contain an Idea which necessarily carries with it another Idea, which latter is one of a pair of opposite Ideas. *E.g.*, The number three is not opposite to the number two, or to the Idea evenness, or to anything at all. But the number three contains the triad and also necessarily thereby the Idea oddness, which latter is opposed to the Idea evenness. (This is the reason why three can never be even.)

6. What in a number causes it to be odd? Not

alone the Idea oddness, but also the monad or the triad, etc., which necessarily contains the Idea oddness. What, then, similarly in a (human) body causes it to be living? Not merely the presence of the Idea life, but also the presence in the body of the soul, which, besides the Idea soul, contains and carries with it the Idea life. Hence, at the approach of death, as three at the approach of evenness, or snow at the approach of heat, soul must either withdraw or perish, for it contains the Idea life, and cannot admit the Idea death; that is, it is undying.

7. Now, if the uneven were imperishable, the number three, on the approach of evenness, could not perish, but would withdraw and disappear. So if the undying thing is imperishable, it cannot perish; it must withdraw on the approach to it of death. Therefore, if the undying thing is imperishable, soul, besides being undying, would also be imperishable.

8. But God and the Idea life and any other immortal thing would be admitted to be incapable of perishing. This indicates that the undying must be imperishable. Hence the soul, because undying, must be supposed to be imperishable, and to withdraw in safety at the onset of death.

On this argument Jowett remarks that it is purely verbal, and is but the expression of an instinctive confidence put into a logical form, "The soul is immortal because it contains a principle of imperishableness." Somewhat similarly another writer (Chase, Bib. Sac. 1849) says that the argument reduced to syllogistic form would be, "Whatever is essentially vital cannot die: the soul is essentially vital; therefore it cannot die," where the major premise is an identical proposition and the minor premise cannot be proved. These opinions seem to indicate that the critics did not get hold of the real point of the argument. That point may be stated somewhat as follows: "Life and death are opposed and incompatible; where either is, the other cannot be. The soul, so far as our knowledge goes, is inseparable from life; it brings life with it, it never leaves it behind when it leaves the body, and it never lingers behind when life is gone; we cannot therefore conceive of the soul apart from life; a dead soul is something outside of human experience, as much so as hot snow or cold fire. Hence we infer that the soul is incapable of death, and as that is the only form of destruction known to us, is immortal." We may represent to ourselves the vital part of the argument by a modern parallel: "Heat implies motion, is an external sign of it, and is inseparable from it. Wherever we perceive heat, we infer motion. Wherever we produce motion, we infer and expect and find heat. So soul and life are uniformly connected in our experience. Wherever we observe life, we infer soul. Wherever we find soul, we may expect to find life, not death." In thus representing the argument, we aim not to go beyond Plato's reasoning, but merely to adapt it to modern forms of thought. There are many criticisms that might be made upon his reasoning. Thus, for instance, he seems to juggle with the word ἀθάνατος, using it now in the sense *deathless*, and again in the sense *imperishable*. Again, it is to be observed that the argument fails to establish personal immortality. If we recur to the parallel suggested above, we reflect at once that it is possible for the motion to be in one centre and for the heat radiated from that centre to appear in a number of bodies which have no heat or motion of their own. So it might be that life was radiated from a central source of life to a number of souls with absolute universality; and then when a certain time came, it might be re-absorbed by the original source, so that it would no longer belong to the individual soul.

III.

ON PLATO'S SYSTEM OF EDUCATION AS PROPOSED IN THE *REPUBLIC*.

PLATO'S system of education as proposed in the *Republic* is not to be understood as presenting his ideal of intellectual culture for all human minds. This needs to be kept in mind when we think of criticising it; for without remembering this, we should be likely to do him injustice in comparing his scheme with those which have prevailed in other times. The modern theory of universal education, for instance, rests on totally different aims, and therefore contains totally different principles from his. But this special restriction in his case does not forbid comparison of his scheme with others; it only compels a candid critic to use greater caution. There is enough of common matter in almost all systems to furnish ground for comparison.

The first thing, then, to notice in his scheme is that it is designed solely for his ruling class. We might easily overlook this fact when we read his remarks on the use of the poets which come in under the head of παιδεία (*education*) in music and gymnastics, II. xvii.–III. xviii. For what he says there seems, as we hastily read it with our ideas, to apply with equal truth to the whole people. But we must observe that it is introduced by the remark that the φύλακες (*guardians of the state*) must be φιλόσοφοι τὴν φύσιν (*philosophers by nature*); therefore, as the παιδεία is one of the means for making them so, it must be intended for them alone. Certainly we cannot suppose that he meant the whole population to deserve the name φιλόσοφοι with his high conception of its meaning. This, then, as well as the more elaborate and advanced part of the scheme in the seventh Book, is designed solely for the small and carefully selected ruling class. And he clearly indicates again and again that it is a difference of natures that determines the selection of some and the rejection of others. Only some specially qualified natures are capable of meeting the tests of fitness for this education, and of these probably some again would be weeded out by the severe discipline of the education itself. It is plain, then, that we must not compare Plato's scheme with general theories of education, which undertake simply to show how the mind can best be developed and instructed. There is another reason why this comparison cannot be made. This scheme is not only for selected natures, but it also has a definite purpose in view. It is the work of a lawgiver, and aims to produce men qualified to do the work of government. Neither of these things is true of what I have called general theories of education. They aim to make scholars, it may be, or cultivated gentlemen, so far as their power extends, but not specially rulers.

If, then, his scheme has a specific object in view, may we not fairly compare it with our systems of special education, those, I mean, of the technical and professional schools? No, we cannot, for the very opposite reason to the previous one. Our general education is, at least in its aims, too general to be compared with Plato's scheme; and, on the other hand, our special educations, for particular lines of work, are too special and limited to be so compared. In the modern theory they are supplementary to the more general scheme, and make no pretension to supply what is supposed to come from it. There is no modern scheme, then, which covers the ground which Plato aimed to cover. If any person attains to such results, it is by favoring circumstances and by work on his own part, of a kind and at times outside of all formal systems.

Can Plato's system be briefly stated? It is set forth in separate parts of his work, in an order determined by the time of life of the pupil. First comes μουσική ("*music*"), including the literature and music which is to form the character from the very earliest youth. He aims to control the nursery stories which mothers and nurses tell to children (*Rep.* 377 C), and proposes to have them, in their representations of the gods, in their heroic examples, and so in their unconscious effect upon character, in harmony with what the young rulers are to hear and believe all their lives. The music, too, allowed in his state is to be such only as will contribute to his main end, and even the metre of poetry must do the same. (This, it may be observed in passing, illustrates not only the sensitiveness of the Greeks to these things, but also the wide reach of Plato's plan, which left no agency unused to influence the development of his selected natures.) Alongside of this mental training, he provides for a bodily training, (γυμναστική, "*gymnastic*") beginning almost as early, and lasting, like the other, through life. Here he does not give quite so full details, but in general outline prescribes a system of simple, harmonious, unremitting exercise, prohibiting all excess, and especially the use of medical treatment to keep life going in spite of sins against laws of health. These two elements of μουσική and γυμναστική form a sort of foundation on which, in the seventh Book, he builds up his advanced education which is to constitute the difference, as I understand him, between the two classes of the φύλακες (*guardians*). They have alike the former training, but this higher education is designed only for those who have shown themselves by his tests worthy to be the rulers. In this, mathematics come first, in the order arithmetic, plane and solid geometry, astronomy, and the science of harmony in sound, which are to be studied however only in theory or, as he expresses it, by problems. After mathematics come dialectics, by which we may understand logic in a wide sense, the science of reasoning, or the laws of thought. Apparently the time for the mathematical training is the ten years from twenty to thirty, and the next five years are to be given to dialectics; then fifteen years are to be spent in the active duties of civil and military government; and from fifty years of age on, the man or woman is to be contemplating the Idea of the Good, and controlling the state in its highest concerns.

In looking at this scheme of education, one thing that strikes us is that its two parts are, or seem to be, controlled by distinct and different ideas. In the first part, the leading idea is a moral one; the aim to be attained and by which the methods are determined is a moral aim. This is plain in the treatment of literature. Nothing is to be admitted, no matter how great the name or the skill of any author, which will give to the youth of this ideal state wrong ideas of the character of the gods, fear of death, or license in excessive indulgence of any emotion. So also as to musical modes and rhythms; such as are simple and severe in their moral effect are alone tolerated. So again as to gymnastics, in the wide sense of all the treatment of the body. Everything is to be done which will contribute to the production in the trained person of εὐλογία, εὐρρυθμία (*beauty of language and of rhythm*), and all the other compounds of εὖ. Even medicine must be watched and disciplined, to see that it does not in any way pander to vice and weakliness, and help men to evade their penalties. In a word, as Plato says (*Rep.* 410 C), the teachers of both μουσική and γυμναστική have in view the improvement of the soul. But when we come to the second part of the scheme, in the seventh Book, a different principle controls everything. In both mathematics and dialectics, the aim of every study and of the way in which it is pursued throughout, is to conduct the mind to the contemplation of real existence, τὸ ὄν, and especially to the highest and brightest part of real existence, the Idea of the Good. This conception of real existence controls, for instance, his mathematical method, and explains why he will have only pure theoretical geometry and astronomy pursued, no surveying, no observation of the heavens; for all external objects are but images or shadows of reality, and only turn the eye away from the only existing thing — the pure Idea. I have said that these two parts of the proposed education seem to be governed by two different principles; but it may more truly be said that the principle is the same in the two cases and the difference is only in the form. For it would be difficult not to see, in Plato's "Idea of the Good," the highest conceivable existence; in other words, it is his name for God. There can be nothing beyond this, as he describes it in the sixth Book; and if the principle of the second part of his scheme of education is the effort to turn the mind to the continual and intelligent contemplation of this divine reality, must we not admit that it would tend to the best possible moral results? Many things in the earlier part point forward to this, — the elevation of the conception of the gods of mythology, by requiring absolute truth in the description of them, insisting on simplicity and reality, and excluding all mimicry. The apparent inconsistence between the two parts diminishes when we remember that with Sokrates, and with Plato, too, virtue is explained as the knowledge of what is right and best. They could not always maintain and carry through its consequences such a theory, — no man can, — but in their reasonings they uphold it. With this as a prior conception (that it is only from ignorance or blindness of mind that men do wrong), it is easy to see how these mathematical and logical studies, pursued, as Plato proposes, to the end of attaining the knowledge of reality, would be introduced into a scheme of education having a moral aim. Thus we see that he uses, in describing the effect of his proposed education, precisely the term which religion has adopted from the Bible, the "conversion of the soul." How different an idea this is from that in the usual Greek word παιδεία, or the Latin word *educatio!*

When we look at Plato's scheme in comparison with our modern schemes, we notice several points of difference in matters where they may fairly be compared. Plato's scheme extends to body as well as mind. This has never, I believe, been a part of the educational system of any nation, unless it has become so of late years here; that is, though the European universities have teachers of riding, fencing, *etc.*, connected with them, the attendance of pupils is entirely voluntary, and the connection merely nominal. But the state, in a rude way, does what Plato suggests when it requires of every man a term of service in the militia. Plato's scheme covers the whole life of the man, keeping him, as it were, in college from twenty to thirty; in a higher professional school five years longer; then again, after fifteen years of discipline in places of responsibility, calling him back to pursue still further the study which was the climax of his earlier education. This may remind us that we have here the work not of one charged with the organization of a system of education, nor yet of a legislator in an actual state. Neither of these men could venture on such absolute control of the lives of men from beginning to end. This is merely an ideal, and the ideal character of it appears perhaps as clearly in this feature as anywhere else. We notice again what seems to us a notable want, in the entire absence of the historical and natural sciences. There is nothing said of languages, history, political science, or, on the other hand, of mechanics, optics, or zoölogy. The general reason for this omission is plain: these sciences were hardly in existence, we should say; but more precisely, they were not yet so developed as to become part of the common property of educated men. Back of this, of course, is another reason, which may be most concisely stated in this form — the comparative absence of books. Books were not easily and rapidly multiplied; the reading class was very small; the conception of book education on a wide variety of subjects for any but the few was not formed. Thus, although, in some of the subjects named, certain individuals had in Plato's day made great progress, particularly in political science and in some branches of natural science, yet the idea that some degree of theoretical knowledge of them, a mere smattering if you will, belonged to the education of youth, had not occurred to anybody. Plato's idea goes far beyond the usual education of an Athenian boy of his time, but does not include this side of the modern idea. It might be added that Plato, though himself a wide student, had a bent towards metaphysics, which would keep him from recognizing fully the claims of the physical sciences. The linguistic science, if it deserves the name of science, of his day is the object of his ridicule in another dialogue; and history, though the type of writing it had been fixed for all time by Herodotus and Thukydides, had not reached a form in which it could be taught. It has sometimes been asked how the average of Athenian education would compare with the modern average, or how the cultivated Athenian gentleman would compare with one who would deserve such a description in our time; and in answering, it is properly said that the works of art in constant sight, the dramatic exhibitions and public recitations, the speeches in the assembly and in courts, must have made up an education which would not suffer greatly in the comparison. This is true, but we must remember that the educational influence of these things was not their prime object, but an incidental result. It may be, however, that the fact of such an effect was one reason why no more elaborate system of intentional education was organized. The men who would have been the ones to see the need of it, and plan it, and keep it going, were also the very ones who appreciated the effect of the influences above mentioned on their countrymen. Perikles's funeral oration in the second book of Thukydides shows this clearly. Demosthenes, a century later, makes a similar remark.

How far does Plato, in constructing this scheme, draw upon his own experience, or how far does it correspond with what we know of his own education? There seems no reason to doubt that the early part of his life was passed like that of other young Athenians of good family. The story that he was inclined to make of himself a poet has no improbability about it, and is indeed confirmed in a measure by the strong evidence in his writings of poetic taste and genius. His writings, too, show very clearly that he was well acquainted with, and sensitive to, the influence of the poetical literature of his people. Nor was there anything in the circumstances of the state during his youth, in the first two-thirds of the Peloponnesian war, to prevent his growing up under quiet influences, as at other times. We are told by a tradition that when he was twenty years old he first met Sokrates, and was drawn away, by the fascination of his society, from every thought of other pursuits. He attached himself to his new master, undoubtedly for the remaining seven or eight years of the life of Sokrates, and after his death was for several years absent from Athens, studying the philosophic ideas of others, and developing his own system. There is special note in tradition of his meeting Archytas, the noted Pythagorean mathematician, in Tarentum. It appears certain that he was strongly influenced by the doctrines of the school of Pythagoras, and particularly by the mathematical element in them. On his return to Athens, he became head of a company of students of philosophy, and remained there for most of his remaining years, elaborating his system and writing his later dialogues. Now, it seems natural to see in this outline of his life something of a resemblance to the plan for his ideal rulers. First, the usual study of literature and of the arts of poetry and music, with a gymnastic training of the body, which latter no one can doubt that he himself had in youth. Then the taking up of severer studies, wholly in the line of mathematics; then the final devotion to metaphysics. May we not reasonably account, in this way, for his choice of these two subjects, mathematics and metaphysics, for the food and exercise of his selected minds, from the fact that he had found his own path of mental growth to lie through them, and in this order? I should think we might venture to say that another thinker, who had followed a different course himself, would probably have marked out a different one for his ideal state. And it may be added that Aristotle, whose course of education, in part with

Plato, was different from Plato's, has left a scheme of education which, so far as it goes, is very unlike the one in the *Republic*.

IV.

THE *OEDIPUS REX* OF SOPHOKLES.[1]

A WORD of apology may be allowed me at the outset. I am quite aware of the apparent audacity of coming to speak upon the subject I have chosen, in this city and in this room. For you have made this play in some sense your own, and there are scholars here far better qualified than I to expound its meaning. Besides, in this room, — if these walls could speak, they might reproduce the thrilling tones of the actors and the chorus of last May, and you can hardly look upon this stage without having brought back vividly to memory those striking combinations, the beautiful group of suppliants, the dignified chorus, the impassioned Oedipus, the graceful form of Iokasta, and all the other elements of the admirable reproduction, — a memory which will make any words of mine seem tame and feeble. But I remember that a certain one also of your own poets, in prose not less graceful than his verse, likened himself in the opening of an address before the Phi Beta Kappa in 1870, to the humble mechanic who goes round, when a train stops at a station, with lantern and hammer, to test the soundness of the wheels. In

[1] Lecture given in the Sanders Theatre, at Cambridge, before the Harvard Philological Society, April 26, 1882.

distant imitation of his example, I would compare what I hope to do to-night to the work of a much humbler ministrant, the ignorant boy, perhaps, who lights the street-lamps, or the brakeman who tells you the name of the next station. If, by often going over this road, I am able to name the things that will attract your attention, or if, after the cunning toil of others is done, I can by a mere unskilful touch throw a little light on your path, it will be as much as I ought to aim at.

The *Oedipus Rex* of Sophocles is one of the most interesting and most distinctly characterized of the extant Greek tragedies. Though it does not contain, like the *Prometheus*, any profound intellectual conception of permanent significance and value, nor any character of terrible majesty in crime, like the *Agamemnon*, nor yet any pure and noble heroine, like the *Antigone* and the *Electra*, it still deserves to rank with these great poems, as of kindred though different excellence. In elaboration of plot, in the complete and sustained presentment of a natural story, it has no superior among the Greek plays preserved to our time. If we accept Aristotle's definition, or rather description, of tragedy, that it excites fear and pity, and thereby purifies the soul in the sphere of such emotions, we can hardly find a better illustration than this play furnishes, to help us understand the description clearly. For here the pity and the fear which a sensitive reader feels are centred on the same person; their causes are no conscious relations or intelligent actions of his; and the character which made him liable to such suffering is, by the very same elements, such as to attract our sympathy. A sketch of the course of the action will perhaps make this manifest, and will serve as an introduction to some comments upon it.

The play opens with the visit of a company of the priests and people of Thebes to the palace of their king, Oedipus, to entreat him to find them some relief from the pestilence which is desolating the city. He has been in peaceful and prosperous possession of the throne for perhaps ten or twelve years, although he did not come to it in ordinary succession. His predecessor, Laios, was killed by some person or persons unknown while on a journey away from home, and at nearly the same time Oedipus, coming as a stranger to Thebes and guessing the riddle of the Sphinx, was rewarded with the throne and the wife (Iokasta) of the missing king. Four children had been born to them, and their life had been one of undisturbed happiness until the coming of this pestilence. Gratitude for that former deliverance, and affection to him as a loved and trusted ruler, naturally bring the suppliants to Oedipus in this new trouble. They describe the sufferings of the people, and appeal to him, almost as to a god, by his previous succor, to help them now again. Oedipus in his answer declares, as would be expected, that the woes of the people were known and keenly felt by him, and tells them that he has already sent Kreon, Iokasta's brother, to inquire of the oracle at Delphi how the plague could be checked, and that he was then looking for his return. As he utters these words, the priest who had been the spokesman of the rest sees Kreon approaching with his head crowned with laurel, which is interpreted as a sign of good news. Kreon comes upon the stage, and announces in answer to the questions of Oedipus that the oracle declares the plague to be due to a pollution of the land by the presence in it of the murderer of Laios, and that it could be checked only by his banishment or death. This leads to a series of questions from Oedipus in regard to the murder, of which he knows nothing; and Kreon in his answer states that one of the companions of Laios who had escaped reported that he was killed by robbers who met the party in the highway and slew all but himself, and adds that the investigation of the matter at the time had been prevented by the all-absorbing distress occasioned by the presence of the Sphinx. Oedipus, forming at once the theory that the murderer had been some one bribed by a party in Thebes hostile to Laios, declares that he will do all in his power to discover and punish the criminal as a mere measure of self-defence, lest a similar plot should be formed against him. Thereupon, at his suggestion, the suppliants retire, having accomplished their purpose, and willing to leave the matter now with the king and the god who sent the oracle. Here ends the prologue or opening act.

The chorus, consisting of elderly men, citizens of Thebes and representatives of its people, now comes forward, apparently summoned by a messenger from Oedipus. In his presence, but before he has spoken to them, they break out in a prayer to the gods for help in the city's trouble. They describe the distress arising from the plague in similar terms to those already used by the priest, and found upon it a yet more urgent appeal to Zeus, Apollo, Artemis, and Bacchus. They know that an oracle has come, but what it is they know not; hence they can only pray in vague terms for relief.

Oedipus, in response to their prayer, states to them the proclamation which he proposes to make, and on which he seems to rely for the discovery of the criminal more than upon prayer. It calls upon whoever has any knowledge in regard to the murderer to communicate it at once to him, and threatens a sort of banishment, or rather excommunication, upon him who hides his knowledge. The chorus, accepting his adjuration, deny all knowledge of the matter, and suggest that the prophet Teiresias should be consulted. Oedipus has already sent for him, and, as he now comes in, proceeds to inform him of the oracle, and asks his help to discover at whom it points. To his surprise and indignation, Teiresias refuses to give any information, saying he would not have come if he had fully understood the purpose for which he was summoned. Oedipus urges him, but he persistently refuses, asserting that he, Oedipus, knows not what he is asking, and that for his sake it cannot be told. In the excitement of dispute, Oedipus at last charges the prophet with having been himself privy to the killing of Laios. Then Teiresias is roused to charge upon him, at first very vaguely, but with growing clearness, that he, the king, is involved in the pollution and guilt which has brought such disaster on the country. In the violent altercation which follows, Teiresias refers for confirmation of his words to Apollo, whose minister he is, the god of prophets and oracles. This instantly reminds Oedipus that Kreon had just come from the Delphic oracle of Apollo, and suggests to him that Kreon and Teiresias were in conspiracy to eject him from the throne. This idea, in harmony with his previous theory of a former plot against Laios, takes firm possession of his mind, and he expands it in terms of bitter reproach. Teiresias is stung by this attack into more express revelations of the condition in which Oedipus is now ignorantly placed, and the terrible future that awaits him; but Oedipus, blinded by anger, and misled by his fixed theory of the motive of the prophet, cannot understand him. A chance allusion on the part of Teiresias to the parents of Oedipus arrests his attention, and makes him ask a question, which, if plainly answered, would bring out the whole truth; but Teiresias is too angry, and only tells him he will soon learn what he asks. Then, with another enigmatical threat, he leaves the stage, and the second act ends.

The chorus, having now learnt the answer of the oracle, wonders who the guilty man can be, yet feel sure that it is hopeless for him to try to escape the punishment which the gods are preparing for him. The altercation between the king and the prophet plunges them into perplexity and distress, for they regard both men with confidence and respect, and cannot tell which is in the right. Yet they decide, for the present, not to give up their faith in Oedipus, who has shown himself, on thorough trial, such a benefactor to Thebes.

Kreon now appears, having heard the rumor of charges made against him by Oedipus, and eager to clear himself from them. Oedipus presently comes out from the palace and states plainly the accusation of conspiracy to get possession of the throne. Kreon, of course, denies the charge, and proves the entire absence of any reasonable foundation for it. But Oedipus is not convinced, for he has conceived his own theory of the matter, and will not readily give it up. He declares his purpose to put Kreon to death as a necessary measure of defence for the state and for himself. At this point, Iokasta, attracted by the sound of their voices in high dispute, comes out and remonstrates with them for thus wrangling in public. Both men address themselves to her, Kreon with an appeal to the gods asserting his innocence. She calls upon Oedipus to respect that oath, and is supported by the chorus to the same effect. Oedipus yields to their urging, so far as to let Kreon go safely away, but does not yet lay aside his anger. When Kreon is gone, Iokasta asks the cause of the quarrel, and learns from Oedipus his theory, that Kreon had instigated the prophet to denounce him as the murderer of Laios. To relieve him from any anxiety arising from prophets or oracles, she tells him then of a previous prophecy, in regard to the death of Laios, which had been falsified by the result. It had been foretold that he should die by the hand of a son of himself and Iokasta: but they exposed their only child to die on a mountain, and Laios was long afterwards killed by robbers, at a place where three ways met. So, she reasons, there is no use in paying any heed to prophecies, if they are not sure to be fulfilled; the will of the gods is better declared by the results they bring to pass. Her story was meant to comfort Oedipus, but one phrase in it disturbs him rather. He asks more particularly about the place "where three ways met," and learns that it was in Phokis, not far from Delphi. Then he asks about the time of the killing, and is told that it was just before he himself came to Thebes. His interest increases, and he inquires what sort of a man Laios was in appearance, and in what company he was travelling when he was killed. The answers make him still more agitated, and he insists that the man who had escaped to tell the story, and who was now a herdsman at a pasture far from the city, be at once sent for. Iokasta promises this, but naturally asks, in her turn, why he is so much excited by her answers. He then tells her, what strangely, perhaps, he seems never to have told her before, the story of his life up to his appearance at Thebes, — how his father was ruler of Korinth, and he had grown up respected there until one day a man at a feast insulted him with the charge that he was not really born of his supposed parents. When he appealed to them, they resented the intimation, but, as it still rankled in his mind, he finally went off secretly to consult the oracle at Delphi. There he got no information as to the past, but a terrible statement as to the future, — that it was his destiny to slay his own father and to be joined in incestuous marriage with his mother. In dread of such a complication, he wandered off, careless whither he went provided it was not back to Korinth, where were the only father and mother he knew. As he walked along the road, he met a party with a chariot, and, becoming involved in a quarrel with them, killed, as he supposed, all of them. The place, — " where three ways met," — the time, the appearance of the leader of the party, the number of persons in it, all correspond with what Iokasta has told him of the circumstances of the killing of Laios, so that he greatly fears lest he may be himself the man guilty of that crime and under the curse of excommunication that he has himself pronounced. Must he be an outcast again, still unable to return to Korinth lest he may unwittingly fulfill that terrible oracle given him at Delphi? He only waits to see the man who had escaped and brought news of the murder, to learn whether he will say that Laios was killed in conflict with a single robber or with several;

for, if the latter, he is certainly clear. Iokasta still encourages him, for the earlier oracle had said Laios was to die by the hand of his own son, and that certainly had not proved true, so that he need not be so afraid of oracles. "True enough," says he; "but still send for the herdsman." Here ends the third act.

Now comes in, to interrupt the course of action, the second song of the chorus. They are even more distressed and perplexed about the matter now, and do not seem so sure as before that justice will speedily be done. The coincidences which disturbed Oedipus in his confidence do not seem to have fallen upon their loyal minds with so much force; but the impious contempt for oracles expressed by Iokasta shocks them; their song is a prayer and a protest against such sinful daring. They will not cease to make the god their defence. There is some dreadful mystery in this violation of the eternal laws of heaven, a fearful outgrowth of pride and excess. If such deeds are to go unpunished, where is religion and the honor of the gods?

As if in answer to their prayer, Iokasta comes out from the palace in a very different frame of mind from that with which she had gone in. Oedipus has been aroused and excited by what she had told him, beyond her power to understand or control him, and in a kind of panic she comes to supplicate the very god whose oracles she had spurned, to help her now. To her in this temper comes a messenger, who seems to bring just what she was wishing for. He comes from Korinth to say that the people there desire Oedipus to be their ruler, since Polybos has just died from old age. She sees at once how much this means, forgets all about her prayer to Apollo, utters in a word her regained scorn of the oracles, and eagerly sends for Oedipus to tell him of the death, from natural cause, of the man whom it had been foretold that he should kill. It seems too clear a case for him to doubt any more, and so he joins in and even outdoes her contempt for the falsified oracle. Yet there is one thing that makes him hesitate to go at once to Korinth and accept the throne, — the wife of Polybos is still living, and there was something in the oracle about his marrying his mother, which may somehow come true so long as she lives. When he gives this explanation to the messenger, he laughs at such a fear, and, with the single purpose of clearing it away, tells him that he is not the true son of Polybos, but that he himself, the messenger, had once, when a messenger on Mt. Kithaeron, received him as an infant from one of the shepherds of Laios, and had given him to the king of Korinth to bring up. So if Oedipus cares to trace his real descent, he must find that Theban shepherd; and he turns to Iokasta to inquire about him, whom the chorus think to have been also the attendant of Laios on his last journey. But she has heard too much already. She tries to turn off the question carelessly, as of no importance. When he persists, she implores him by the gods, by his love of life, by his love for her, to forget all that has been said and abandon the whole inquiry. He mistakes her motive, thinking she fears he may find himself to have sprung from a low family, and, now thoroughly aroused to solve the old long-forgotten doubt as to his parentage, determines to follow up this clue and find out, at any cost, who his parents were. She leaves the stage, in silent agony of despair, foreseeing the terrible revelation. He remains, despising her woman's pride, and trusting that the good luck which has given him this throne, and whose child he jestingly calls himself, will still befriend him. So the fourth act ends.

The chorus, taking his tone, rejoice in the thought that soon his mysterious parents, perhaps some mountain nymph and wandering god, will be made known, and all the trouble ended; but their song is short. The old Theban shepherd comes in, sent for by Iokasta, we must remember, as the only man who had witnessed the killing of Laios; but there is no thought now of asking him about that. He is confronted with the Korinthian messenger and recognized by him at once. His own memory is feebler, but with a reminder from him he recalls their old acquaintance. Then he is asked about the infant, and told that Oedipus, king of Thebes, is the same person. At once he suspects what is coming, and refuses to answer any questions. By threats from the king he is compelled to tell what he knows, and so the dreadful truth comes out that Oedipus himself is the son of Laios and Iokasta, and already the murderer of his father and husband of his own mother. Oedipus curses himself and rushes into the palace.

In bitter contrast to their last hymn of joy and hope, the chorus now bewail the lot of man, so brief in its enjoyment of prosperity, as this example teaches with terrible plainness. They are still loyal to their former regard for Oedipus, and have no feeling towards him but pity.

Then comes a messenger from within the palace, and narrates, according to the custom of the Greek stage, the dreadful events that had occurred within. He tells how Iokasta had come in, tearing her hair and lamenting bitterly, and had entered her chamber, when suddenly his attention was drawn away by the entrance of Oedipus, raving and calling for a sword, and demanding to be told where Iokasta was. No one would tell him; but he suspecting, burst open the doors of the chamber and there found her hanging lifeless. Then, with most furious curses, he snatched the golden buckles from her dress, and with them tore out his own eyes, — that they might nevermore, even in Hades, see the persons involved in his unwitting crimes. While the chorus is lamenting his madness, he comes forth in his wretched blindness, carrying the buckles still in his hand, and after incoherent exclamations to himself, recognizes the voices of his friends, and joins them in bewailing his misery. Then, in a long passage of somewhat calmer tone, he justifies his self-mutilation, and reviews his life in the light of the shocking discovery of his real relations to those about him. While he is thus speaking Kreon comes in, whom he knows not how to address, remembering those unfounded suspicions which he had held when he was mentally blind. But Kreon re-assures him in kind words of pity, and presently has his two daughters brought out to him, whom Oedipus entrusts to his care, bidding them a tender farewell. Kreon then leads him away, with some last words which hint that he may not continue to be so gentle and friendly in his treatment of the helpless sufferer; and the play ends with the reflection from the chorus that no man can be pronounced happy until his life is seen through to its last day.

This outline, inadequate as it must seem to one who knows the original, may yet be of use in recalling the distinguishing peculiarity of this play, viz., the degree to which its interest depends upon the plot. It is the only one of the Greek tragedies, with perhaps a single exception, in which a secret is kept from most of the persons concerned until near the end, upon which secret the whole story hangs. In the nature of the material used by the Greek tragic poets, it was almost impossible that this should often be the case. For they used old myths which were familiar to the audience in their whole structure, and in which, therefore, it was not easy to succeed with a surprise. It should be said, however, in justice to their art, that this was not always a hampering restriction. They constructed their plays in free recognition of it, making the interest and the exercise of poetic and dramatic power depend upon other elements than intricacy of plot. They allowed themselves, too, some measure of freedom in the treatment of the traditional myths, in minor points,— a freedom which was abused by the last of the three whose works we have, whether from lack of invention or from some defect in his principles of art. It is a signal triumph of the strong and disciplined genius of Sophokles, that he constructs this play with a catastrophe perfectly familiar to his audience, yet so skilfully that one might hear it often and still be as much absorbed in the unfolding action as if he were as ignorant of the end as the characters are supposed to be. Let us analyze this delicate work a little.

The poet has two objects to accomplish in laying out his plan. One is to bring about the revelation of the secret of the birth of Oedipus in a perfectly natural way, without the voluntary intervention of any human agent. This absence of voluntary human agency is emphasized by the poet, and seems to have been necessary, in his view; perhaps for the reason that he wished to show how the gods work out their plans without the conscious help of man, and even against his will. The means which the poet uses to bring about the revelation are the plague, the quick temper of Oedipus, the death of Polybos from old age, the hope of gain on the part of the Korinthian shepherd, and the love of the Theban shepherd for Laios. The plague lies at the foundation of the whole action, so far as it is contained in this play, furnishing not only a natural opening scene (the point in dramatic art which was such a stumbling-block to Euripides); but also the condition without which the succeeding events could not, in their present shape, be explained. It is the first shock to the prosperity of Oedipus since he came to Thebes, and, of course, it soon brings a strain upon the weak point in his present position. It is a blow from the gods, aimed directly at him, in such form that, while it reveals nothing to him, it compels him to act, and, by his action, to bring out at last the whole secret. He acts promptly in the way which seems to him best at the time, and which yet recoils upon him later. The moment his proposed course of action receives a check from Teiresias, his temper is roused, and he becomes committed to a theory which he holds obstinately. This theory makes him quarrel violently with Kreon, all without suspicion that he is preparing repentance and woe for himself. The quarrel with Kreon brings Iokasta on the scene, and she, merely to relieve him from anxiety regarding the alleged oracle, tells the story of the death of Laios, in which one incidental phrase, "the place where three ways met," gives the first serious shock to his conviction of innocence of the murder. This leads to a review of his life, and so brings back to his thought the never-solved question as to his parentage. All this grows naturally, and without intention, out of his quickness of temper. Then comes in the shepherd from Korinth, whose hope of a reward (as is stated in lines 1005 f.) had made him eager to be the first to bring the news to Oedipus of the throne awaiting him. It seems strange that he should be the very man who had received him when an outcast in infancy; but, apparently, the poet supposed that he was stimulated by that hope of gain to keep himself informed as to the life of Oedipus after he left Korinth. His revelation that Oedipus was not the son of Polybos comes out in simple, ignorant desire to deliver him from anxiety about returning to Korinth, not from any purpose to contribute to the exposure of his hidden calamity. Finally, the Theban shepherd, the one person in the country, unless we except Teiresias, who knew the murderer of Laios, and the only one who knew that the son of Laios and Iokasta might still be living (though, of course, even he had never connected the two things), comes in to do his part. At his own request, and apparently from a love to Laios that made the sight of his murderer on his throne intolerable, he had been sent out of the way of telling his knowledge since Oedipus came to Thebes, and had heard nothing of the new oracle about the plague. He is now brought in to testify as to the murder, and, against his will, is compelled to testify as to the parentage. Thus it appears that every incident, except the plague and the oracle, comes into the series by human action, from some motive entirely apart from the discovery of the guilt of Oedipus.

The other object of the poet is to put Oedipus in the wrong in his attitude towards the gods in this part of his life, so that his terrible fate may not seem wholly undeserved. This is a matter which it is important to have fully recognized, if it is true, because it is not apt to be recognized at all in modern estimates of the play. Most of the current popular references to this story speak of it as one in which a perfectly innocent person is dragged by a cruel fate, determined for him before he was born, into horrible deeds, and then into dreadful, unmerited ruin. The representation of the play here a year ago furnished the occasion for a vigorous article in a Boston periodical, based wholly on this false idea. It so happens that, besides the evidence in the play itself, we can bring an independent, ancient authority of no little weight to prove the falseness of that idea. It is well known that Aristotle, in his treatise on poetics, uses this particular play perhaps more than any other, to furnish illustrations or proofs of the rules he lays down. One of these rules is, that the hero of a tragedy must be a noble character, but not without ἁμαρτία, that is, not without some fault or defect; on the ground that, if he is a perfectly innocent person, his suffering would not excite the spectator's pity or terror, but rather, his indignation and horror; and if, on the other hand, he is made too great a villain, the spectator would merely think he was getting his deserts. Then he goes on to mention two examples of such noble heroes with the requisite faults, and one of them is Oedipus. Now the mere dictum of Aristotle is not one to which every head must bow, and it may be that some wholly admirable tragedy has been written with a faultless hero. But the thing for which I quote him is his opinion that, in fact, Oedipus is not made by Sophokles such a faultless hero, and on that point, surely, his judgment ought to be respected. Let us then see, if we can, in what the fault of Oedipus consists. We find that the course of the action brings him, by virtue of his own character and conduct, into such a relation to the gods as cannot help suggesting to a Greek audience some painful result. His first words to the chorus, after their prayer for divine relief, convey a hint that he is disposed to trust more to his own proclamation and the authority of the government than to the help of the gods. Then his sudden anger and wanton suspicion in regard to the prophet, and contempt for his sacred character, would seem the very thing to draw down upon him some punishment. When Iokasta first expresses her disregard of the oracle, he does not interrupt her with rebuke. When she gives reasons, and speaks yet more scornfully, he assents. When, finally, the messenger comes with news of the death of Polybos from old age, which seems to put beyond all possibility the fulfilment of the oracle that he should kill his father, he rivals her in triumphing over the baffled prophecy. These things are not forgotten. When, at the end, he comes out blinded and bleeding, and the chorus ask him what god has impelled him to such a deed, he answers, "Apollo! Apollo it was who brought to pass this bitter, bitter woe of mine!" But Apollo was the god of the Delphic oracle. Though, indeed, the ultimate cause of his misery was his involuntary parricide and incest, yet the shock with which the discovery came, and its fearful consequences, are to be ascribed to the sin of contempt of the gods, into which too great confidence in his prosperity had betrayed him. No one can study the tragedies of Aeschylus and Sophokles without recognizing the prominence in their view of this particular kind of sin, as provoking the wrath of the gods. In this play it takes a form corresponding to the rest of the plot, and seems to be brought on inevitably by the action of such incidents upon such a character.

One or two minor features of the plot also deserve passing notice. The fact that it was Kreon, the natural successor to the throne as regent, in the minority of the sons of Oedipus, who was sent to Delphi for information as to the plague, prepares the way for the suspicion that he was in league with the prophet to put Oedipus out of the way. The quarrel with Kreon subsequently not only furnishes an occasion for Iokasta, his sister, as well as the wife of Oedipus, to come out and remonstrate with them, thereby bringing her with her story of the murder into the action, but also adds greatly to the impression of the closing dialogue between the two men in relations so changed,—Oedipus repentant, and Kreon forgiving. This last scene would have been just as possible, though not nearly so effective, if, in the previous scene, Oedipus had merely expressed to others, in Kreon's absence, his suspicions of him. The introduction of Teiresias likewise effects a twofold purpose. On the one hand, his refusal to answer the question put to him starts Oedipus on his course of opposition to the gods, and, on the other, it is his relation to Apollo, the god of Delphi, that suggests collusion on his part with Kreon, and thus introduces the train of events that follow upon the quarrel with the latter.

The whole play, like many others, is marked by, or rather consists of, a series of alternate movements in opposite directions, and with opposite effects on the feelings of the audience. First, the deputation of priests describe the sufferings of the city under the plague; and then Oedipus, comforting them with sympathy, is presently enabled, by Kreon's arrival, to point out a definite cause of the calamity, and to promise that every means shall be taken to remove it. Next, the chorus, on its entrance, fills the mind again with the dismal scenes of the general misery; and Oedipus again, by his strong, confident declarations of what he is going to do, seems to clear away half the trouble at once. The strange conduct and incredible statements of Teiresias cannot fail to make the hearer dread something, though he hardly knows what, before which Oedipus seems helpless as a child; but when he has to deal with Kreon, though we may think him hasty and overbearing, yet he seems strong enough to crush mere human opposition, and to make a way for the State into peace. In his conversation with Iokasta, however, he is plainly overpowered by the close coincidences of her account of the murder with his own recollections, and feels again the presence of some mystery which may be too much for him. The coming of the news from Korinth naturally lifts him into freedom from fear, but it is only for a moment; and the determination which the other fact, learnt from the same man, excites in him to discover at any cost his real parents, presently plunges him, in spite of the gleam of hope seen in the song of the chorus, into the depth of misery. After the first rush of horror and self-condemnation, there comes again a reaction, and the play leaves the audience at last somewhat soothed by the comparatively quiet final scene. It is manifest how these changes add to the life and interest of the action, and also how they serve to retard the movement of events, and postpone the coming of the fatal discovery. Some one has said of the *Odyssey* that the whole plot would be broken down by the existence of a post-office system, so that Penelope might have heard from Odysseus occasionally. Surely, in this play, if Teiresias had come to Oedipus in a calm hour, and told him what he knew as a prophet about his life, or if, by any other natural means, he might have learned it earlier, the whole structure would break down. However, it is plain that the poet, with artistic design, makes a gradual approach to his climax, letting the anger of Oedipus prevent his believing, or even listening to, the significant words of the prophet; making Iokasta do all she can to dissuade him from pursuing the investigation after she sees whither it tends; bringing in the news from Korinth to give him a moment's delusive comfort before his fall. A notable instance of this designed delay has been already mentioned, — that when Teiresias says, "Your parents, however, thought me inspired," Oedipus suddenly asks him, "Who? Stop! Who were my parents?" An answer to this question would have ended the play there, but Teiresias has been angered beyond such compliance, and puts him off with the riddle, "This day shall bring you parents and ruin."

It is in some sense a consequence of this character of the plot that the play exhibits in especial frequency what has been called the *irony* of Sophokles. The word *irony*, though perhaps the best that our language affords, does not strictly in its English use express the idea that is here intended. If I understand its modern usage, it implies generally some measure of contempt, good-natured contempt sometimes, when a man feels perfectly sure of his own position or powers and plays with an adversary, but still a real looking down upon one who might claim to be an equal. The Greek word, as defined and illustrated originally, does not seem to have implied this. The quality is defined by Aristotle as the pretence or assumed appearance of being worse in some respect than one really is. He speaks of irony as lying on one side of the truth, and of boasting or arrogance as lying on the opposite side. And he acutely adds, that the pretence of the worse or humbler condition may proceed from something very like arrogance; which recalls the story that Diogenes, in his squalor, walked in over the rich rugs in Plato's house, saying, "Thus I trample on the pride of Plato." "Yes," answered Plato, "and with a no less pride of your own." The prime illustration of irony in this Greek sense is Sokrates in the dialogues of Plato, where he assumes the tone of ignorance and desire for information, and through his questions exposes the ignorance of another.

Now the word *irony* as used of the dramatic poet means something different from either of these senses. For the poet has no adversary, and cannot properly manifest contempt for his characters. He is the creator of his mimic world, and so acts a part toward it like that of the divine governor of the real world. Hence, in this use of the word, it means the same as when we speak of the irony of fate. The tragic poet, deeply feeling the pathetic contrasts that arise in the development of his story, and knowing that the audience will feel them too, chooses to set them forth most forcibly by showing the hero in his glory just before his utter ruin, or in his apparent humiliation just before his triumph, and by making the character say in his unconsciousness what has a different meaning or a deeper meaning than he can yet suspect. I have been thus minute in speaking of the different senses of the word *irony*, because a recent editor of Sophokles, Professor Campbell of St. Andrews, has objected to the use of the word to describe a quality of the Sophoklean tragedy, on the ground that the offensive sense of superiority, the sneer of contempt, which belongs to the word in its ordinary use, is out of place in the relation of the poet to his characters. That is quite true. No such thing as a comparison between the poet and the character on the stage, to the disparagement of the latter, can be imagined. But Professor Campbell does not sufficiently recognize the other use of the word, as in the phrase, *the irony of fate*. That phrase justifies the application of the word to the work of the dramatic poet, for he is in a sense the Fate of his characters, the author of all that they say and do. From him proceeds the practical irony, the conflict in the dramatic situation between the reality and appearance, and the verbal irony, — that is, the putting into the mouth of a character words that would seem to a person so situated to be true, which yet have a pathetic force of contrast from the knowledge on the part of the audience that, as the speaker means them, they are terribly far from the truth. The practical irony of this play has been admirably expounded by the late Bishop Thirlwall, in his essay on the Irony of Sophokles, and I will pass it over. The examples of verbal irony are of course mainly to be found in words put into the mouth of

Oedipus. In the very opening of the play he describes himself as "renowned in universal fame." In declaring his purpose to obey the oracle with zealous effort to detect and punish the murderer of Laios, he says, having in mind his theory that it had been prompted by hostility to the government, that self-interest prompts him to "put such villany far off from himself," for a similar attack might be made upon him. In his proclamation calling for information, and denouncing any one who shall withhold it, he adds at the end, "And if he be any inmate of my house, the curse applies to him as well." When Iokasta urges him not to try to discover his parentage, he ascribes her entreaties to a fear that he may turn out to be of a low family, and says to her (his real mother), "Not even if my mother was a slave, and her family in slavery for three generations back, will you be degraded by it." A similar conflict between reality and appearance is seen in the language of the chorus, especially in that brief ode just before the revelation of the secret, when, like children playing on the edge of a precipice, they amuse themselves with conjecture as to what god and nymph it may have been, who in some wanton hour begat him who had come to be their king. These are but a few examples of an element which pervades the whole play.

It has been said that the main interest of this play lies in the skilful treatment of the plot, but the remark should not be understood as implying that the characters are in themselves feeble or uninteresting.

Oedipus ranks high among the figures of Greek fiction, and he owes his position wholly to Sophokles, and largely to this play. He is himself his only enemy. Every other character in the play is friendly to him, and strives to help him. His very strength becomes a cause of weakness and calamity to him, under the circumstances in which the gods place him, because it betrays him into self-confidence, and blinds him for so long a time to the truth. From the first it is evident that he is a man of strong will and clear head. The people of Thebes, after long experience, reverence him as their ruler. The people of Korinth, who had not seen him since his youth, send for him at once, on the death of Polybos, to take the throne. There is no feebleness or indecision, either in his action when he is taunted with being a foundling, when he hears that threatening oracle at Delphi, and when he meets the party of Laios on the highway, or in his words at the beginning of this play. He has already sent Kreon to Delphi, and as soon as the response comes back, after a few pointed questions in regard to the crime, just such as a modern police magistrate might ask, he has his plan formed, and makes his proclamation. He is full of self-reliance and energy. The opposition of Teiresias only fixes his purpose more firmly, and he makes up his mind at once to deal with Kreon as the offense he imputes to him demands, without thought of fear or favor. When his attention is again drawn to the unsolved question of his birth, he pushes the inquiry in that direction with the same energy, in spite of all the entreaties of Iokasta. And when he learns the bitter truth, his vengeance upon himself is no less sudden, severe, and appropriate. After the blow, how clear is his inward vision over his past life, how complete his self-subjection! It is thus evident that his very clear-sightedness for what lies just before him, and his promptness of action, are what bring upon him, so far as his deeds affect his fate, his faults and misfortunes. They make him act too quickly and confide too much in his own judgments. Yet, on the other hand, he is conceived as a man who made his way everywhere, and attracted to himself the love and respect of those around him. The language of the chorus, as well after as before his fall, shows this plainly. We might liken him to Achilles, the ideal warrior of the Iliad, — impetuous, truth-loving, self-impelled rather than self-controlled, capable of feeling and arousing in others intense affection, and hardly less intense hatred, keenly sensitive to the judgments of others upon his conduct, yet, under the influence of excited passion, adopting a course for himself in defiance of all around him, and persisting in it in defiance of reason, at terrible cost to himself. Or, to take an example more widely known, from a period of history not very unlike the fabulous heroic age of Greece, he was such a man as David, the partisan chief, the hero of outcasts, the king of Israel, the poet, the sinner, the penitent.

The character of Iokasta, too, though subordinate and less fully drawn out, is worthy of study. The idea of the poet embodied in it, to be inferred from the words he puts into her mouth, seems to have been often misunderstood. Nearly all who have referred especially to her, regard her chiefly as an impersonation of impious disbelief in the gods. Thus Campbell calls her "the arch-horror of the piece." Capellman, in his essay on the Women of Sophokles, finds something admirable in all the others, but has hardly a word to say in her favor. Schneidewin, — a most judicious critic generally, judging a poet's work with a delicate and cultivated tact, — describes her as selfish and heartless, unconcerned at the death of Laios, careless what became of the child maimed by him and exposed by her, indifferent to gods and oracles alike, until she finds herself driven to heed them by terror and distress. Now, if Sophokles had imagined her such a person, would he not have drawn the picture so clearly and strongly that there would have been no room for doubt or difficulty in receiving the impression? Yet, if one reads the scenes in which she appears, and the references to her, with this question in mind, — In what light did the poet himself look upon her character? — he will hardly come to such a conclusion. Her first appearance certainly impresses one in her favor. When she comes out to Oedipus and Kreon, at the height of their wrangling, both men defer to her at once, with great respect, and state their cases to her. She then, with the aid of the chorus, brings them to a settlement, or at least a postponement of their quarrel. This shows that she was not a weak character, and creates a presumption that she was not a wicked one. As to her attitude towards the oracles, it ought not to be overlooked that she repeatedly distinguishes between the direct manifestations of the divine will, and the possibly mistaken or falsified declaration of it through human channels. Such a distinction, we may be sure, was made by many not undevout people in the poet's time. The Delphic oracle itself was accused of having Medized in the terrible trial of the Persian invasion. Other instances of suspected tampering with its utterances are mentioned. And so the questioning of the genuineness or supposed application or suggested fulfilment of an oracle was probably no uncommon thing. In other plays of Sophokles the possible defeat of a prediction of evil is part of the plot. That the chorus here is shocked at the apparent impiety of such distrust does not prove that every one, even the poet himself, if he had been treating a different myth, must feel so. Moreover, in this case, she had seen, as she fairly argues, one such instance in her own experience, — Laios, as she thought, had not been killed by his own son, but by robbers on the highway. And it is to be observed that there is no hint of her having shown any distrust of oracles, until she finds that her husband is angry with the prophet Teiresias for charging him with a murder which he is sure he did not commit.

This leads us to what seems to be the central and ruling quality of the character, as here drawn. There is in the Iokasta of Sophokles no more prominent trait than her love as wife for her husband. We may indeed guess — for we are told nothing about it — that it was love to Laios that led her to consent to the exposing of the child, because that seemed the only way to save the father from death by his son's hand. When, after the death of Laios, she is given by the State to its deliverer from the Sphinx, she comes under the influence of his character, and after a time so loves him as to cast contempt on the oracle for his sake. When she finds that she cannot thus allay his anxiety about the killing of Laios, her conscience distresses her, and she appeals for help to the very god whose Delphic oracle she had scorned. And at the last, when the secret of the birth of Oedipus becomes known to her, while yet unrevealed to him, her first thought is to save him from the dreadful discovery. She is willing to try to keep it to herself, to live on with that fearful secret torturing her soul, if only she can secure for him the bliss of ignorance. It is the blind impulse of unreasoning love — precisely such a one as the same poet represents in the case of Ismene, when she urges Antigone to accept her as a partner in death, by falsely admitting that she had been a partner in the burying of their brother — it is, I say, a blind impulse of unreasoning love, for such a secret could not long be kept by her or hidden from him; but she is carried away by it, without stopping to think what it means and involves. When he persists, she has nothing left to her but suicide. Now I appeal to you who are familiar with the play, whether such an interpretation of her words and deeds is not more natural and true than one which makes her out a cold, heartless, skeptical fiend?

This play suggests a question which is worth asking, for the sake of the view it opens into the work of the poet in such a case: Could Oedipus have avoided his fate by any wisdom or effort of virtue? Certainly he need not have killed Laios. The story of the collision on the highway, as he tells it, does not imply any attack upon him by the other party which justified his violence as an act of self-defense. If he had yielded the way to the larger party, — he a mere foot-passenger, and their wagon, perhaps, running in the deep-worn grooves in the rock-bed of the road, such as are to be seen now in parts of Greece, — there would have been no such fatal result. Again, he might have refused marriage under any and all circumstances, to ensure the failure of the other part of the oracle given him at Delphi. It appears thus that forewarned as he was by that oracle, it lay within his power, by such careful self-restraint, to prevent its fulfilment. It was only a mistake of judgment in supposing that he knew whom the oracle meant as his father and mother, that betrayed him into realizing its prediction. Had his idea been right, his precaution of not returning to Korinth would have saved him. And so it is simply a confirmation of the conception of his character already suggested; for his ruin came from over-confidence in this opinion of his own. In many ancient stories it is such a problem on which all the interest hangs: Will a man, forewarned of an impending calamity, be able by foresight, caution, wit, or daring, to defeat the purpose of the gods and avert or evade the calamity? And always, in the story, there is some point where his knowledge, or self-control, or watchfulness fails, and the will of the gods is done. We must bear in mind that it is just such a story that Sophokles has here dramatized, and that he must take the main incidents as he finds them, without material change. It would, in fact, destroy the story to give it a different issue. The original myth may have had a very different meaning. Indeed, there is not a little probability in the theory that the germ of it is one statement that the day destroys the night from which it sprang; and another, that the sun, after much wandering, returns at evening to the beautiful twilight, from which at morning he came forth. When the meaning of the terms for the daylight and the sun was lost from memory, so that they became proper names to the ear, as Zeus and Selene and Aurora and many others did, as Grace and George and Augustus and all the others have done more recently, the old statements became narratives of supposed human action instead of descriptions of natural phenomena, and so a story grew out of them.

All the rest, the oracles, the collision in the highway, the guessing of the Sphinx's riddle, the children, the suicide, and the self-mutilation, were engrafted upon the original stock to supply motives, or natural consequences, for the human action. But, however that may be, the fact remains that Sophokles took the story as he found it, for the basis of his dramatic treatment; and the question as to his work, the test of his skill, is this: Is the story, in his version of it, in all respects such a story as might be believed to occur in the heroic age of Greece? It is not possible to prove an affirmative or negative answer by comparison of actual events, for we have no record of facts from that time. The only special material available for an opinion is the picture presented in the Homeric poems and in the other tragedies, which aim to represent in the main the same social state. If we judge this play in the light of these works, and on such general principles of human nature as are true in all ages, we find it a natural, self-consistent story. Given a man born under such a fate, led by an unseen control through such an early life, and the rest of the life, as here developed, presents nothing unnatural, nothing out of the range of human experience. It is a marvelous story, and the supernatural element in it is essential to the structure; but such an element is recognized in the belief of all ages, and here it nowhere interferes with the action of ordinary human motives and emotions. It simply avails itself of these springs of human action, and brings the persons into such relations to one another that their natural conduct in these circumstances produces these momentous results. Oedipus may act in this or that particular as one or another of us would not act, but all we can ask of the poet is that Oedipus should not ever act otherwise than as the conception of his character would require; in other words, that he should be consistent with himself. The mysterious Sphinx (who also is perhaps a personification of a natural phenomenon) and the plague with which the play opens are, besides the oracles, the only elements which connect the story with fairy-land or the supernatural world. The special development which our poet gives to the bare outline of the myth, the incidents which were necessarily introduced to fill up a story of human action on the basis of the phrases describing phenomena of external nature, will be found to lie entirely within ordinary human life in the social state here depicted.

In what has been already said of the dramatic skill displayed in the plot of this play, it may seem that too much has been claimed for the Greek poet. It might easily be that a reader familiar with Shakspeare, or with almost any dramatic poet of the modern era, would think in going through this play that nothing was done in it, that there was no action, but rather, an excessive amount of talk. This difference we need not try to deny or to apologize for; but it may be in part explained by considering certain restrictions under which the ancient poet did his work. One was in regard to the number of actors. In Greek tragedy (not in comedy) the highest number allowed by rules of usage was four, and most of the plays preserved to us could be acted by two or three persons. Another was the restriction of unity of time and place,— that the whole action should be confined to a single day and to one locality. These unities were not, it is true, strictly regarded by the Greek poets, for the first is violated in the *Agamemnon* of Aeschylus, and both in the *Eumenides*. Yet there was some force in them, and Sophokles has observed them in all of his plays that are preserved to us. Once more, there is the rule stated thus by Horace in the *Ars Poetica:* —

> *Ne pueros coram populo Medea trucidet,*
> *Aut humana palam coquat exta nefarius Atreus,*
> *Aut in avem Procne vertatur, Cadmus in anguem;*
> *Quodcumque ostendis mihi sic, incredulus odi.*

This explains the comparative absence from the stage of conflicts, suicides, transformations, etc., and the introduction of long narratives of messengers which constitute in part the epic element of Greek tragedy. These, and other restrictions, are little matters in themselves, but they would greatly hamper a modern playwright. They all belong indeed to a higher cause, arising from the essentially different conceptions of ancient and modern tragedy. The Greek tragedy was in origin, and in theory always, a chorus interrupted by dialogue. The chorus was at first the whole performance, and the dialogue passages, called *epeisodia* from the entrance of the actors to take part in them, were truly episodes in the sense which the word has to our minds. Hence it was only by one at a time, in the course of years, that the number of allowed actors was raised to three or four. This explains the comparative absence of scenery and action from the stage. And hence, too, perhaps, from the relative subordination of the actor's part in the play came its limitation to the single place and time. The whole tragedy was a poem in illustration and explanation of a series of *tableaux vivants*. It may fairly be said in view of these restrictions, and of this theory of tragedy, that Sophokles has in this play shown wonderful power in developing a complete and absorbing plot. In comparison with other plays it seems as if he here strained the Greek conception of tragedy to its utmost limits in a direction approaching the modern conception. And yet how differently a modern writer would treat the theme! He would have three or four times as many characters. He would have a second or third subordinate plot; and it would go hard with him if he could not work in a love-story with reasonable obstacles to its running smoothly. He would omit the heaven-inflicted plague, and transform the blind old prophet into a prime minister or a ghost. He would bring about the discovery of the fatal secret by some chain of half-accidental occurrences, like the dropping of Desdemona's handkerchief, such as might occur in every-day life. He would change the scene a dozen times, and lengthen the time of the action indefinitely. In saying this, of course one does not mean that the modern dramatic form is necessarily inferior on account of these differences. It is merely that we have here two distinct conceptions of this form of art, the ancient and the modern, or, if you please, the statuesque and picturesque, though this latter word has been in bad company and lost some of its native simplicity. Each form is the best in its own age and surroundings. Each in comparison with the other appears to have weak points, but has, not less truly, strong points peculiar to itself. The remarkable thing in this play is, that the poet without being false to the classical conception, has been able to introduce so much of what characterizes more especially the modern form of dramatic art, — an interest in the mere series of incidents, and a probable secret naturally brought to light.

It may be worth while here to point out certain improbabilities which appear to a modern judgment in the story, as presented in this play. One is, that it should be so long before the plague, or whatever declared the divine displeasure on account of the killing of Laios, came upon Thebes. Oedipus must have been in undisturbed possession of the throne for years, since he has four children born to him by Iokasta, the age of whom at the time of the action is not, to be sure, expressly stated; but from the last scene in which the two girls are brought upon the stage, one gets the impression that they were at least no longer infants. And there are expressions here and there in the play (vs. 109, 561, 1212) which contribute to suggest a long interval since the coming of Oedipus to Thebes. The murder of Agamemnon, it is true, remained unavenged for seven years, but that interval was necessary to the story, in order that Orestes, who was but a child when his father went to the ten-years' siege of Troy, might grow to sufficient age to be able to avenge him. Here there appears no need in the story for a longer time than that these children should be begotten, and it is worth notice that, in the brief Homeric version of the myth, no such interval before the discovery appears. Another strange thing is, that the death of Laios, known to be a murder on the highway, should have been passed over with so little notice. The poet himself felt this difficulty, and suggests, as an explanation of it, that the distress occasioned by the Sphinx had interrupted a search for the murderer, which was not afterwards resumed. Of course, in an unsettled state of civilization, and among a group of small, independent states, such acts of violence were more likely to occur and to defy punishment, and generally, in the primitive societies, homicide was a less serious offense, as appears from recognized tariffs of payment in money for it to the outraged family. But still it remains a strange thing that the king should be so taken off with no more serious and prolonged investigation of the matter. This, it may be observed, is an inseparable part of the original story, and does not belong only to the dramatic working up of it. Again, one cannot help asking how it happened that the same man should have been a shepherd on Mt. Kithaeron at the time of the birth of Oedipus, then an attendant of Laios on his fatal journey towards Delphi, and afterwards a shepherd again. This last change is expressly accounted for by the poet, as caused by the man's desire to get away from the sight of his master's murderer on his master's throne. It has been suggested above that the poet's own reason for thus removing this man from Thebes was to exclude the possibility of his revealing, during the reign of Oedipus, his knowledge that he was the murderer of the former king; and besides, the delay in the plot occasioned by the necessity of sending to some distance for him contributes to the suspense and interest of the spectator. But there is no hint of an explanation how it came about that the same man was a witness of the exposure of the child and of the killing of the father. It is easy to see how necessary it was to the plot, as Sophokles conceived it, to have the same man cognizant of both events, though ignorant that the infant and the homicide were the same person. The whole identification depends upon his testimony.

Finally, can we say, after all this, what was the poet's motive or aim in this play? There can be no doubt that he had some conception in his mind, some definite motive, which controlled the shaping of this creation. Without such conception, the work would have little meaning or value. How distinctly it was present to his mind and formulated in expression, we cannot guess, but we may be sure that such an artist as Sophokles did no work at random. In the selection of a myth for dramatic use, and of the precise version and part of the myth, we may suppose that he would be guided almost entirely by his perception of dramatic possibilities; at that earlier stage of the process, the creative faculty of the poet is not yet at work, or is at work only tentatively and fitfully; his mind is rather passive, receiving propositions, as it were, considering and comparing, but not yet acting upon any. In this stage, the artistic element predominates over the true poetic (*making*) element, and the mind, with comparative coolness, selects for artistic reasons without determining the moral and drift of its future work. But when the work of composition begins, and the fire burns within, then the whole man gives shape to the product; his long-cherished thoughts, his beliefs about the highest and the deepest questions, his principles of action, his noblest theories, — for into such work the true poet will put the best of everything, — all these will be poured into the crucible, and will give something of form or color to the final result. And it is not presumption for any one to attempt to discover from the finished work what is the ruling idea, the main thought, in it. The poet speaks to his hearers and readers. He is not a juggler, aiming to distract our attention with by-play, and to hide his real design, but a teacher, whose object is to convey, in such form as shall suit him best, his mind's thought to other minds. We may fail from our own weakness to read his thought rightly, but our honest effort to discover it is the proper tribute to his effort to convey it. And if we fail, our failure may lead another to the truth.

The story of Thebes seems to have been particularly attractive to the mind of Sophokles. It is commonly supposed, on various grounds, that he wrote first the *Antigone*, taking up that part of the myth which comes last in the order of events, perhaps from the desire to depict, as in the parallel case of the *Elektra*, a heroic woman in a moment of extreme trial. Next probably in order of writing came the present play. We are not wrong, I think, in supposing that this play interested him so deeply in the character and fate of Oedipus that it did not wholly satisfy him, that his mind recurred to it and dwelt upon it until he felt an impulse to treat it once more; and then, in his old age, as tradition tells us, he wrote the *Oedipus at Kolonos*. Now if this semi-traditional order of the three plays is correct, it seems to lead us towards the answer to our question, What was his main idea in this play? He was not drawn to the myth at first by the desire to tell the story of Oedipus in dramatic form. It was not the intricacy of the plot, the exercise of dramatic skill in the natural unforced bringing to light of a secret that first attracted him to the Theban myth, but the strong and pure character of Antigone. The power of her character over his mind led him then to go back and take up the dark story over which her self-sacrifice sheds its gracious light. Perhaps in later years, and in fuller mastery of the resources of his art, he really had more pleasure in the construction of such a plot as this; but that alone does not seem to explain the vigor and passion of the play. If he was led to select the theme by the plot alone, he was soon carried beyond the source of interest by the deeper questions it aroused in him. Let us turn now towards the other Oedipus-play. What was it that he was dissatisfied with in the *Oedipus Rex?* What were the deeper questions started there, and not fully or not rightly answered, to which the *Oedipus at Kolonos* was meant to give the best answer the poet could find? Here is a noble character, — strong, sagacious, religious, — forced to pass through the deepest misery and disgrace. How did it come about? And beyond that, how can we believe in a divine government of the world if such things come to pass under it? To answer these two questions was the poet's object, the same which Milton proposed to himself in the beginning of *Paradise Lost*, to

"assert eternal Providence,
And justify the ways of God to men."

The *Oedipus Rex* has for its object to "assert eternal Providence." As clearly as the poet can, he shows in it how certainly the gods govern in the life of this world, how not even the strongest and wisest and best of men can put his own life outside of the chain of cause and effect, and nullify the decrees made known to men in divine oracles. Far-seeing and firm and devout as Oedipus is, in an unguarded moment he does commit the very sin that is needed to bring him into the fatal sequence,— and all the rest follows without any violent intervention, by the working of ordinary laws. But the poet could not bear to leave the matter finally here. With all his care to show Oedipus to be in the wrong, the impression of undeserved suffering remains. He must go on, as Milton must, and "justify the ways of God to men"; and this he does in the *Oedipus at Kolonos*, in a way which makes us wonder at the depth and tenderness and truth of Greek theology in his hands. In a word, then, we may truly say that the main idea of the *Oedipus at Kolonos* is to show, by an extreme and striking example, how,— again in spite of all appearances to the contrary,— the same divine will and law is able, as soon as man submits to it, to lead him even through bitter suffering into joy and peace.

V.

SUMMARY OF THE *OEDIPUS AT KOLONOS* OF SOPHOKLES.

THE scene of the play is laid at Kolonos, one of the rural demes of Attika, about four miles north or north-west from the Acropolis. When the stage is disclosed to view, we see two persons walking on the public road, — an old man, blind, and in beggar's rags, and a young woman guiding and half-supporting his steps. In the first few words of their conversation they announce themselves to be Oedipus and Antigone. They have been long journeying thus, in search of the place of his final rest and release from the burden of life. They do not know just where they are, though Antigone sees, in the distance, the walls of a city which she knows, from directions given them along their way, to be Athens. The old man sits down at the roadside, on the low wall of an enclosed grove, while she proposes to go and find out where they are. But before she can do this it is made unnecessary by the approach of a man, a wayfarer like themselves, — not a native, apparently, nor resident of Kolonos, — to whom they apply for the information they want. He is horrified at seeing Oedipus on consecrated ground where it is forbidden to go, — the sacred precinct of the Eumenides. Yet he does not venture to compel him to move, and while thus in doubt describes to him the region to which he has come. The whole region, the stranger says, is holy ground. It belongs, in general, to Poseidon, but Prometheus, the giver of fire, has in it a place sacred to him; the grove is a sanctuary of the Eumenides, called the corner-stone of Athens, and the surrounding land is under the protection of its eponymous hero, the equestrian Kolonos. Politically, the whole territory is under the government of Theseus, king of Athens. Having said thus much, the stranger advises Oedipus to remain where he is, while he goes to inform the men of the deme of his presence there, that they may decide what he shall do.

This passage gives us a glimpse of the character of an Attic deme. It is a distinct community, gathered in one locality, like a village in a New England town. It has its own gods and sanctuaries, gods who may also be worshipped elsewhere, or may be peculiar to that spot. Several deities of different characters may divide among them the reverence and worship of the little community, each having his own enclosure and temple. It has a measure of self-government, as to its own affairs, so that it might expel Oedipus from its limits; but, at the same time, it belongs to a larger body, and recognizes the authority of Athens over all Attic territory.

After the stranger is gone, Oedipus utters a prayer to the dread Eumenides, imploring them to fulfil the oracle of Apollo, which had promised him that he should find rest at the sanctuary of some dread deities, though *where* it did not tell. Now, since, without intention or knowledge, he had stopped first at their threshold, on coming into Attica, — he, a sober man, at the door of deities who abhorred all use of wine, — it seemed as if he must have been guided thither in the divine plan, and this must be the place meant by the oracle. So he fearlessly throws himself on their pity. At the close of his prayer, Antigone warns him of the approach of some elderly men; whereupon he withdraws into the wood, that he may learn in what temper they come before he shows himself.

The chorus announced by Antigone comes forward in great excitement, eager to find the wanderer who has profaned the sacred enclosure. While they are urging each other to look everywhere about the grove for him, Oedipus calls out and discovers himself. Something in their words may have encouraged him to this, but rather, perhaps, he sees that he cannot long evade their search, and thinks it wiser to give himself up. They are filled with awe and pity at the sight of him, but, nevertheless, they insist upon his coming out at once from the sacred grove before they will talk with him. Then follows a passage in which the poet strikingly depicts the hesitation, timidity, and physical weakness of the old man. "Daughter, what shall I decide to do?" "Father, we must do as the citizens here do."

"Give me then your hand." "Here it is." "Friends, let me not be wronged if I trust you, and leave the protection of this sanctuary." "No one shall force you away from this place." "Further still must I go?" "Come on." "Further yet?" "Lead him forward, maiden, for you can see." And so it goes on, while he groans over the trouble he has in getting to the spot they indicate. The hesitation and helplessness of Oedipus here, in this trifling matter of walking a few steps, is in strong contrast with the courage and resolution he shows later in the play, when upon his decision rests the fate of two of the chief states of Greece. Here it is bodily action that is required of him; there it is an act of the mind, a decision to be made and maintained. Here he is still uncertain of his position, whether he will be suffered to remain in Attika, or must wander further; there he had learnt from the king that he may stay.

When he has reached the spot designated for him by the chorus, they proceed to ascertain by urgent questions who he is. He resists as long as he can, but at last, by the advice of Antigone, tells them his name. Knowing something of his terrible story, they are horrified at learning that it is Oedipus who is before them, and instantly bid him quit their territory. He reminds them of their pledge, but in vain. They insist that he deceived them (by not telling his name, apparently), and therefore they are not bound by their promise to him. Thereupon Antigone appeals to them to have pity upon her, as they might upon one of their own daughters, and not to drive her and her father away. They remain unmoved; the fear, so general among primitive peoples and in ethnical religions, lest the whole community may suffer from the sin of one of its members, or for harboring an offender against the gods, is too strong to give way readily. Then Oedipus himself addresses them, taking a higher tone than before, and not only asking as a favor, but claiming almost as a right, shelter in Attika: What will become of the reputation of Athens as a most religious city if she casts off this suppliant, a man more sinned against than sinning, whose evil deeds as men regard them were wrought in ignorance, who comes now to Athens as a man consecrated and bringing a blessing with him? What he means by this blessing he will explain when the ruler of the land appears. The chorus is awed by his words, and consents to his remaining until the king comes, adding that the same man who summoned them, the stranger who first came upon Oedipus, had gone on to carry the news to Theseus, and they feel sure that he will soon be there.

Their conversation is interrupted by an exclamation of surprise from Antigone. In answer to an anxious question from her father, she tells him that she sees approaching a woman mounted on an Aitnaian steed, with a Thessalian hat to protect her head from the sun. (Those who have seen the Tanagra figurines will recall the shade-hats which appear on some of them.) Can it be? Yes, as she comes nearer, she sees her smile, and recognizes beyond a doubt her sister Ismene. 'The description of the comforts with which Ismene travels is evidently designed to mark by contrast the hardships which Antigone cheerfully undergoes, to which her father presently alludes.' After they have exchanged affectionate greetings, Ismene says she has come with news for her father. "But where are my sons?" he naturally asks. "They are where they are, and there is trouble between them." This gives occasion to Oedipus to denounce the conduct of his sons in staying at home regardless of his fate, and to contrast it with the love of his daughters, one of whom has borne all the hardships of his wandering and want with him, and the other has come now this second time to bring him information. What, then, is her news this time? It is that the two brothers have quarrelled about the throne of Thebes; that Polyneikes, the elder, has been driven into exile, and that, according to the prevalent rumor, he has found friends in Argos, and will presently come with an army to regain his rights. Then, in answer to questions from Oedipus, it comes out by degrees that the oracle at Delphi has lately made known to the people of Thebes that the possession or control of Oedipus's grave was essential to the welfare of Thebes, although, as a parricide, he could not be buried in Theban soil. It appears from what he has previously said, that substantially the same thing had before been told by the oracle to Oedipus himself, only in the somewhat different form that his grave should be a blessing to whatever land should contain it. In consequence of this oracle, she further tells him, Kreon is coming soon to get possession of his person, in order that they may keep him close to the borders of Theban territory so long as he lives, and bury him there when he dies. He asks whether his sons knew of this oracle, and when told that they did, and yet set the possession of the throne before any care for him, breaks out into curses upon them, enumerating their misdeeds towards him, and praying that their strife with each other may never end.

The more the chorus sees of Oedipus, the more favorably inclined towards him they become, and now, as if regarding his remaining in Attika as a settled thing, they call his attention to the ceremony of purification necessary to propitiate the Eumenides, upon whose sacred soil he has unwittingly intruded. They describe with minute detail the process, of which he is evidently entirely ignorant. The suppliant must take fresh water from a flowing source in vessels wreathed with wool from a young sheep, and standing with his face to the East, pour three libations of water and honey, without wine; then he must take thrice nine twigs of the olive in his hands, and utter the formula of prayer, including a reference to the name Eumenides (this in a low tone), and then withdraw without looking behind him. This minute account of the ceremony, more minute than we find elsewhere, illustrates the difference of religious usage in different communities, to which reference has already been made. Oedipus, brought up at Korinth, and having lived afterwards at Thebes, both within a day's journey of Athens, has no knowledge of this ceremony, though the deities to be propitiated seem from his earlier words (vs. 99–106) to have been known by him. For there, having been told only their name, he speaks of them as averse to wine, and calls them daughters of Skotos. The preciseness of the directions given, and the eager attention which he pays to each detail, illustrate the importance of such formalities in a religion like that of the Greeks. The use of any other material than lamb's wool for the fillets, or of water from a still pool, or of twenty-four twigs instead of twenty-seven, might vitiate the whole process. Oedipus himself cannot go to perform this rite, nor is he willing to be left alone in his blindness; so he says one of his daughters must go in his stead, and Ismene volunteers to do it. We see then here one of the reasons why she was introduced into the action of the play. But why the poet introduced here this mention of the purification, and so made it necessary to have some one go to perform it, we cannot perhaps be so sure, though we may conjecture.

After Ismene is gone, the chorus extracts from Oedipus by close and persistent questioning, a confession of the dreadful facts in his past, which he cannot mention or hear mentioned without great distress, — the murder of his father, and the union with his own mother. But he insists that both deeds were done in utter ignorance, and without thought of evil on his part. This brief passage seems to be introduced here in order to bring before us yet again his freedom from guilt in the matter, and the horror with which he looks back upon it all. These things are impressed upon us by frequent repetition through the play, and it seems necessary that they should be, that we may understand the favor with which the gods at last regard him.

At this point Theseus, king of Athens, comes in, and the action of the play takes a new turn, the second main incident beginning here. The first main incident is the application of Oedipus for shelter in Attika; the second is his actual reception by the highest authority of the state. The king greets Oedipus by name, after briefly explaining how he has made up his mind that the mysterious wanderer must be he, and asks what request he has to make. Oedipus tells him that he brings his own body as a gift to Athens, and that if the gift is accepted, it will prove a great benefit to the state. When Theseus asks how, Oedipus at first puts him off by saying that time will show; but presently in answer to further questions it comes out that, if Theseus will give his body burial, and resist all attempts of the Thebans to get him away before or after his death, then in a war which shall arise between Athens and Thebes, the Thebans shall be defeated at his grave. Theseus now formally consents to his remaining in Attika, and gives him the choice whether he will remain where he is or go with him to Athens. He, of course, remembering the oracle, decides to remain at the grove of the Eumenides, and Theseus, after pledging to him protection against any attack, goes away.

As if to ratify this promise of the king, and to show what joys are implied in it, the chorus breaks out into a well-known and exquisite song in praise of Kolonos and of Attika. A rough version will give the run of thought. First strophe: "Thou hast come, wanderer, to the choicest region of this land, the white hill of Kolonos, which above all others the nightingale frequents, warbling plaintively among green thickets, honoring the dark ivy and the deity's sacred grove, rich in fruits, which never the sun nor blast of any storm penetrates; where the reveller Dionysos strays with his divine attendants." Antistrophe: "And by the rain from heaven is ever fostered the narcissus, time-honored garland of the two great goddesses, and the yellow shining crocus. Nor do the sleepless rills from the Kephissos ever fail, but continually they flow over and fertilize with pure water the hollows of the hilly land; which land the Muses do not scorn, nor does Aphrodite with golden reins." Second strophe: "And there is (here) what I do not hear of as belonging to Asia, nor ever growing in the great Dorian peninsula — the native self-propagating tree, that no enemy has dared to visit, which here most abounds, the gray-green wholesome olive, which no warrior young or old shall ever destroy, for the all-seeing eye of Zeus and the keen glance of Athena watch over it." Second antistrophe: "And another most choice glory have I to mention for my native land, the gift of a mighty god, the glory of the horse and of the sea. It is thou, O son of Kronos, lord Poseidon, who hast given her this glory, in that it was here that thou didst first bring the horse under the restraining bit. And the oar, framed for the hand of man, flies and leaps over the water of the sea, keeping pace with the thronging Nereids." You see how the poet passes from praise of the special locality, the place of his own birth, to praise which includes the whole land of Attika. The luxuriant growth of the vines and trees mentioned may have been peculiar to Kolonos, but the culture of the olive, the use of the horse and of the oar, we know were not. No version that I have seen gives any idea of the careful structure of the ode, its balanced clauses, its chosen epithets, its harmony of sound and sense. Each of the first pair of stanzas ends with a brief mention of deities, with whose attributes the earlier part of the verse has some connection. Of the second pair, one is devoted to the praise of the olive and of Athena, the other speaks of the horse and the oar, and of Poseidon who gave them for the use of man, thus recalling the myth of the strife between these two divinities for the possession of Attika. It is one of the most charming choruses in Sophokles, and one of the few passages in classical literature that show a pleasure in the beauties of nature. But a modern reader notices at once that there is no reference in it to what strikes us as a chief part of landscape beauty. Nothing is said of the view of Kolonos, or of the view from it, although the latter is one of the most delightful that the modern traveller can find in the neighborhood of Athens, embracing the whole plain with its olive groves and houses, the Akropolis, and the other hills around it, the more distant encircling mountains, and at one point a broad stretch of the blue sea. There is no recognition of the beauty that appeals to the eye, and through it to the imagination, the beauty of distant outline, of ever-varying color, of combination and suggestion. Instead of this we have an enumeration of the several features that make the place delightful or serve the uses of man and of the divinities that honor it.

After this comes the third main incident of the play, the efforts to remove Oedipus from his sanctuary, and to make him take one side or the other in the impending conflict at Thebes. It takes up some seven hundred lines of the play, but we may pass over most of it briefly; indeed, it seems as if some of the proverbial garrulousness of old age had got control of the poet here. After the splendid chorus, Kreon comes in, and, as Antigone says, speedily puts the brave boasts of its words to the test of action. He begins with a smooth speech, professing to be sent from Thebes to persuade Oedipus to come home and hide away among his kindred the scandal of his life. But Oedipus answers him with so much indignation and contempt, that in the wrangling that follows between the two, Kreon presently throws off his mask, and, after boasting that he has already captured Ismene, directs his attendants to seize and drag off Antigone. This they do in spite of the protests of the chorus. Then encouraged by his success so far, he proceeds to attempt to drag off Oedipus himself, which of course was his real aim from the beginning. But now the chorus shouts so loudly for help that Theseus, who is sacrificing at the altar of Poseidon not far off, hears them, and comes to learn what the matter is. His coming quickly changes the state of things. As soon as he learns what has been done, he sends off from those gathered at the sacrifice soldiers to guard the road by which the girls will naturally be taken on the way to Thebes, and then, after listening to Kreon's defense of his conduct, and a long reply from Oedipus, requires the former to guide him to the place where the girls are.

While they are gone, the chorus, unable on account of their age to join in the pursuit, utter a song having reference to the battle which they suppose will occur. They wish they could be present at one place or another where they imagine it to be going on; they predict victory for their countrymen; they pray to Zeus, Athena, and Apollo to fulfil that prediction. This choral song seems designed merely to fill the gap between the departure and the return of Theseus. Some time must be allowed for the rescue, since it is implied by line 1148 that there was something of a struggle between the two parties. But we see from the brevity of the chorus, — only fifty lines, — that the poet was not careful to make the interval seem long enough for the pursuit, the conflict, and the return; no such realism was required by anything in the Greek artistic sense.

At the end of the chorus, Theseus comes in with the two maidens. They are warmly welcomed by their father, who also pours out his gratitude to their deliverer. And here it is interesting to notice how the poet avoids giving an account of the battle, which he seems to have known the audience would expect, and yet to have preferred for some reason not to give. He makes Oedipus ask Antigone for an account of what had occurred. She refers him to Theseus as the proper person to tell of his own achievements. So he turns to him, and, though he does not in so many words ask for the story, yet he evidently expects it, and Theseus recognizes the unuttered wish, but only to decline gratifying it on the ground that he does not wish to boast of what he has himself done, and what Oedipus can learn about from his daughters. "Besides," says he, "another matter was brought to my notice as I was coming here which needs immediate attention." This announcement diverts the thoughts of all parties from the battle, and Oedipus himself starts and keeps up the inquiry about this new matter. "They tell me," says Theseus, "that a man, not a townsman of yours, but yet a kinsman, has sat down as a suppliant at the altar of Poseidon where I was just now sacrificing." "Who is he?" asks Oedipus. "I do not know. I only know he wants to speak with you, and to have safe conduct away by the way he came." "But who can it be who comes thus?" "Consider whether you have any relative in Argos who might have come with such a request." This is enough for Oedipus, who has heard from Ismene that Polyneikes had found friends in Argos; he will hear no more, and refuses at once to see the man, who must of course be Polyneikes. Theseus remonstrates with him, and Antigone pleads, until at last he yields, and consents to his son's coming.

Again, the interval necessary to allow time for summoning the suppliant is filled up with a choral song, but this time it is a more interesting song than before. The thought of it is suggested by the sight of Oedipus as he sits there, old, blind, and poor, and assailed first by the violence of Kreon, and then by the hardly less hateful petition of Polyneikes. "He who desires length of days nurses folly in his heart. For many days bring one into sorrow and there is no joy in them; and at the end, gloomy death stands waiting for all. Best of all is it never to be born; and next best to die as soon as possible and go whence one came. For after the follies of youth come the woes of life, jealousy, strife, conflicts, slayings; and after all these comes friendless, gloomy old age. In such an old age must live not I alone but also this poor man here, on whose head as on some exposed cliff beat waves of calamity from all sides, from west and east and south and north." This passage is one of the famous expressions in ancient literature of the sense of the weariness and emptiness of human life. "All is vanity and vexation of spirit." It is the more remarkable as coming from a poet who was notably of serene and cheerful temper, and whose life was a long scene of success and happiness until perhaps its very latest years. It has been used, with other passages of similar purport, to show that the Greek religion had nothing in it to satisfy the needs of a thoughtful spirit, or again to prove that old age was necessarily a gloomy and cheerless part of life to the Greeks. It does not, I think, prove either of these things, though they may both be true. I am not sure that we can get at the true explanation of such a tone in Greek literature, but it seems to me to be due merely to natural reaction in the midst of a life of activity and pleasure. We may express the idea under various forms; we may say that the full blaze of light requires some qualification of shadow, or that it was the same feeling that prompted the proverbial presence of the skeleton at the Egyptian banquets; or we may recognize in it the feeling that most of us have at some times in youth, — a perfectly natural and genuine feeling, I think, but crude and transient, — that life is hardly worth going on with, and the world is a poor place after all. I should suppose that the very brightness and gayety of Greek life in general would make such a contrast to be keenly felt and strongly expressed by a sensitive spirit, wherever its eye was caught by any of the inevitable calamities of human destiny which cannot be wholly ignored. If such things were possible in the midst of all this joy and revelry, what is it all worth?

After this chorus, Polyneikes appears and makes the second attempt to move Oedipus from his chosen resting-place. He comes in hesitatingly, evidently in doubt as to his reception, and addresses first his sisters, speaking of his father in the third person. Presently he gets up courage to address his father directly, but failing to get any answer he turns again to his sisters and asks them to help him move his father's will. Antigone encourages him to go on with his appeal, and to expect an answer at the end. So he tells the story of his quarrel with his brother, his exile, and his alliance with Argos; he enumerates the heroes who are engaged with him in the attack upon Thebes, and urges his father to give him the help of his presence. For the oracle, as he understands it, promises victory to the party in that struggle (not in a struggle between Thebes and Athens, as Oedipus has heretofore represented it) which shall have with it the person of the old hero. Oedipus hears him through and then calmly proceeds not only to decline his request, but to curse both of his sons in solemn form, praying that they may die by each other's hand. The solemnity and elaborate fulness with which this curse is uttered and repeated show how prominent and important an element of the story it was. Upon it depends apparently the necessity of that inseparable part of the legend, the meeting of the two brothers in battle and their killing each other. Yet the poet seems plainly to show us in the lines which follow that there was no such necessity of sequence as to hamper the free will of either brother. For in these next lines Antigone pleads with Polyneikes to give up the expedition against Thebes, and thus frustrate that part of the terrible curse. Just so in the *Seven against Thebes* of Aeschylos, the chorus pleads with Eteokles not to put himself in the defense of the city just where he will be sure to meet his brother. In both cases we see that the brothers might have avoided their sad fate, but in both the pride of military honor is too strong. Thus we see that in this case, as I believe in all other cases in Greek tragedy, the calamity of an individual is due, not to a resistless fate, but to some error or sin of his own doing.

Scarcely has Polyneikes withdrawn in dejection and disgrace, when the fourth and last main incident of the play, the passing of Oedipus, begins. It is ushered in by a peal of thunder, the meaning of which Oedipus instantly recognizes. He asks that Theseus be sent for at once. The dialogue between him and Antigone is repeatedly interrupted by short stanzas from the chorus, which describe the repeated thunderings, reveal the excitement into which the chorus is thrown, and must have produced in the audience, by breaking in thus with quick exclamations in impassioned metre, a similar effect of excitement and confusion. Oedipus repeats and urges his desire that Theseus should come, and, at the end of the last choral stanza which calls loudly for him, the king appears. Oedipus at once becomes calm, and with great dignity assumes the direction of matters. He tells Theseus that these peals of thunder are a summons to him to go into the other world. He tells him that he, Theseus, alone must go with him to the appointed spot which, in order to ensure the safety of Athens in conflict with the neighboring States, he must keep secret from every one, only imparting the knowledge to his successor when his own life draws near its end. (Thus the poet ingeniously accounts for the fact that in his day no one knew the spot where Oedipus had died.) Then the old man rises in his blindness and becomes in his turn the leader of the others, his daughters being allowed to accompany him for part of the way. As they go off the stage, the chorus begins its last choral song, which is a prayer to the deities of the lower world to give to Oedipus an easy death and a kindly welcome into their domain.

At the end of this choral song, a messenger appears and gives the chorus an account of the last that was seen of Oedipus. He tells how he led them along to a place which he describes, but by landmarks which no longer exist; a place where apparently there was thought to be an entrance to the lower world. Here he sat down, and stripping off his old rags bade his daughters bring him water for a bath and a libation. When this was done and he had put on other clothes, there was heard a peal of thunder from below the ground, whereupon he began to bid an affectionate farewell to his daughters. At a pause in their weeping over each other, there came an awful voice, calling, "You there, Oedipus! Why delay we so? It is a long time that we are waiting for you." At this he must go; he only lingers to commit the maidens solemnly under an oath to the care of Theseus, and then bids them go away that they may not see what becomes of him. When they have withdrawn and waited a little while, they look back and see Theseus standing there alone, shading his eyes with his hand as if some supernatural sight was before him. They look again presently and see him doing reverence to the earth beneath and at the same time to the heavenly Olympus. And no man to-day, except Theseus, knows any more what became of Oedipus.

After this the rest of the party who had gone with the old man return, and the two maidens utter their sorrow in a long *kommos*, in which the chorus join. At last Theseus bids them stop lest they offend the gods by too protracted lamentation. Antigone, in her blind sorrow, begs him to let them see their father's grave, but he refuses because Oedipus had required him not to show it to any one. She acquiesces then and asks him to send them back to Thebes that, if possible, they may prevent the threatened fatal conflict between their brothers. This he promises to do, and so the play ends.

THE *OEDIPUS AT KOLONOS* OF SOPHOKLES. 141

NOTE. — As to the plot: Really no dramatic element. Skill of poet in working in incidents so as to give as much action as possible to the play. In this like the *Prometheus*. There, after the prologue, we have a motionless figure, approached in various ways with attempts to sway his will. So here, after the reception of Oedipus, two unavailing efforts are made to change his purpose. In fact, he, the central figure, sits still in one place for thirteen hundred lines, from 202 to 1540. Variety of incidents to make up for this: Theseus comes in four times; the purification, the violent proceedings of Kreon, the conflict brought almost before our eyes by the choral song about it, the mysterious summons by the thunder peals.

Natural sequence of incidents, especially at the beginning; accidental meeting with wayfarer; motive of introducing chorus; device for keeping Theseus in the neighborhood (*cf.* 888 with 54 f., 1494 f.).

Reasons for introducing Ismene. She brings the news of Kreon's coming, so that Oedipus is prepared for that. Also, she makes known to him the quarrel between the two brothers, and their knowledge of the oracle. This prepares him to receive Polyneikes as he would wish to do. She also supplies somebody to go and perform the rite of purification at the proper place, and her being there, or on the way, enables Kreon to boast of having already captured her.

It may be noted that there is no subsequent reference to this rite of purification which Ismene was sent to perform. We do not know whether it was done or not before Kreon seized her. Furthermore, it does not appear why he should have seized her as a captive except as a mere wanton outrage to the feelings of Oedipus. For nothing is said that implies any intention on her part to abandon Thebes and join her father in his wandering. Why should she not go back with her steed and attendant to Thebes to live, as she had done once before? The seizure of Antigone took away the sole companion and the eyes of Oedipus, but not so that of Ismene.

Other difficulties of plot that have been noted are of little or no real importance.

Relations of Athens to Thebes implied cannot be satisfactorily explained. Much friendly language, yet a conflict anticipated; perhaps the sheltering of Oedipus ought to be regarded as an unfriendly act. But the myth required it, and perhaps the poet did not concern himself with either political relations or contradictions.

VI.

SUMMARY OF THE *ANTIGONE* OF SOPHOKLES.

THE prologue is a conversation between Antigone and her sister Ismene. It is accounted for in the most natural way: Antigone the freer, more active and wide-awake character, has heard some important news which the quieter Ismene has not heard, and it is news the first hearing of which may probably lead to her committing herself to some action in view of it. So Antigone, having made up her own mind, contrives an interview with her sister alone, early in the morning, and tells her that Kreon has decreed that Polyneikes must be left unburied as a penalty for making war on his native city. Antigone, however, has resolved to bury her brother in spite of this decree, and urges her sister to join her in discharging this religious duty. Ismene is too timid or too prudent to venture such defiance of authority, and strives to persuade Antigone by every argument she can think of not to persist; but it is all in vain, and they part without either's having affected the other's purpose. This prologue gives us in brief the theme of the play, the conflict which constitutes the tragic situation. Ismene represents the general attitude, that of everybody except Antigone, — she disapproves the decree, but feels that she must obey it; she admires Antigone's purpose, but cannot bring herself to make it her own. Thus we see the same event acting differently on two different characters, and so developing them in opposite directions.

When the sisters have withdrawn, the chorus comes in, singing the *parodos*. This is one of the finest choral songs in Sophokles. It is a song of triumph over the deliverance of the city and the repulse of the enemy. It is made up of alternate lyric and anapaestic stanzas. The first and fourth lyric stanzas express the joy of the delivered city, the second and third describe the repulse of the foe. The first and third anapaestic passages allude to the unpatriotic action of Polyneikes and the mutual slaughter of the two brothers, thus mingling thoughts of evil with the general strain of joy; the second celebrates the special intervention of Zeus to punish the pride of the assailants, and the last merely announces the coming in of Kreon. The variation of thought accompanies the change of metre, and the choice of words is such as to express the thought most clearly and precisely, and at the same time with richness of suggestion and ornament. The whole is full of brightness and vigor, in harmony with the sunrise with which it opens.

Kreon comes in as announced, and addresses a speech to the assembled elders, complimenting them for their past loyalty, declaring his purpose to rule with firmness and patriotism, and formally publishing his decree in regard to Polyneikes. The chorus bows to the will of the king, and seems to agree to give its support to the new decree. It asserts its belief, however, that no action will be needed, for no one will be so foolish as to disobey, with the penalty of death before him. Scarcely are the words uttered, when one of the guards appointed by the king to watch the body of Polyneikes and see that no one buries it, comes hurriedly in to tell him that in spite of their watching the deed has been done. Here we have one of the best examples of character-talk in the Greek drama. The man tells everything else before he gets to his real message, describes his own reluctance to come with it, in the dramatic form peculiar to common people, evades the king's questions, and lets his own concern in the matter intrude itself, until the reader is in full sympathy with the king's impatience. Then at last he tells how at sunrise they found the body strown over with dust, and how he was chosen by lot to bring the news. The king is very angry, and utters at once his belief that a party among the people hostile to his rule have bribed some of the guards to do this thing. Then he dismisses the guard with heavy threats of punishment for him and his comrades if they do not detect the criminal.

After this comes a choral song of very different character from the previous one. This belongs to the reflective, philosophic type, and is an excellent example of it. Ignorant who has done this deed just reported, and shocked at the daring shown in it, the chorus breaks out thus: "Many are the things that excite my awe and wonder, but none more so than the nature of man!" Then it enumerates the achievements of man which show the boldness, restlessness, and ingenuity of his spirit; how he has made the stormy sea his pathway, how he makes the earth yield him food, how he ensnares birds, wild beasts, and fishes, and has tamed the horse and the bull; the invention of speech, of laws, of housebuilding, the cure of diseases; how for everything he has some device, except that he cannot escape death. Now all this wisdom, if guided to right ends, is a blessing, but if a man seeks wrong ends by it, it is a curse, — and here evidently they have in mind him who has set at defiance Kreon's decree. To this choral song there is a number of parallels, as to the type; and there are similar passages not in choral form, in which is given a brief history, as it were, of civilization, — notably one in the *Prometheus*, where all the arts of civilized life are ascribed to his gift. It is noteworthy that Sophokles here ascribes to man's daring and inventiveness the very things which elsewhere are regarded as taught by gods or heroes to men. This simply illustrates the absence of fixed systematic doctrine in the Greek religion. Each poet might represent things on each occasion as the occasion demanded; or as his own tradition said, even if it conflicted with other tradition.

At the end of the choral song is an anapaestic stanza

in which the entrance of Antigone under guard as a criminal is announced. Kreon opportunely comes out from the palace at the same moment (why, we are not told), and to him the guard, the same man who had come before, reports with something of the same style as before (unable to leave out of view his own feelings and opinions) that they had caught Antigone in the act of performing burial rites over the body of Polyneikes. In answer to Kreon's question she confesses the deed; thereupon, he dismisses the guard and asks Antigone how she has dared to defy his command. In reply she utters the famous lines avowing a belief in divine law as superior to any human enactment whatever. Thus she justifies her action and declares herself ready to meet the consequences of it. But this plea is of no avail in the eyes of Kreon. His mind is filled, to the exclusion of everything else, with the idea that his decree has been set at naught, and that all opposition to it must be put down by force. He is not at all embarrassed by the proved falseness of his previous theory that the guards had been bribed by disaffected citizens to bury Polyneikes; but he rages against Antigone as if he had all along known that she was the criminal. He includes Ismene, too, in his fury, and, without any reasonable ground of suspicion, sends for her to answer the charge of complicity. Meanwhile the argument between him and Antigone goes on until it is interrupted by an isolated anapaestic stanza from the chorus, announcing the approach of Ismene. She comes in weeping, and is rudely asked by Kreon whether she had a share in the burying of her brother. To our surprise she answers that she had, if Antigone says so. This is one of the delicate touches of the poet in illustrating character. The timid girl, who has urged her bolder sister not to venture such a deed, is now so influenced by the heroic act and critical position of Antigone that she wants to be with her in everything. She could not share her daring before, but she can share her death now. But Antigone, of course, will not consent to this. In the dialogue that follows she seems to us needlessly harsh and cruel to Ismene. Perhaps all we can say about it is that the poet so conceived her character, that in this trying situation, with every nerve held tense in the purpose to meet death in any form without flinching, she would naturally be unable to make allowance for feebler spirits, or to allow herself any moment of tender feeling. Such seems to be her attitude here, and we must admit that the impression her words make is a painful one. Yet they do not chill the affection of Ismene, for, when Kreon interrupts the dialogue of the sisters, she turns to him and pleads, but in vain, for the life of Antigone. She is the first to mention Haemon, the son of Kreon, betrothed to Antigone, and thus the way is prepared for his appearance in the next scene. Kreon orders both the sisters to be led into the house, the proper place, he says, for women.

Then comes the third choral song, of similar type

with the one immediately preceding, yet not quite the same. That was entirely abstract and general in its thought, with no explicit reference to the special occasion; this begins and ends with general reflections, but between them comes a verse applying them to the case in hand. It opens with a text, as it were: " O, blest are they whose lives are free from touch of woe!" Then comes a magnificent simile, in which the successive calamities that befall some doomed families are likened to the billows that sweep on the Aegean sea, driven by a north-east gale from Thrace, and dashed on the shores of Greece, full of sand and sea-weed. The antistrophe sees in the Labdakidae, the royal family of Thebes, a case like this; Labdakos, Laios, Iokasta, Oedipus and his two sons, have all perished miserably, and now Antigone, the last of the race (the existence of Ismene being ignored for the moment), is to be cut off. The second strophe magnifies in noble language the sleepless, immortal, irresistible power of Zeus whose offended law brings on these calamities. Yet not without the sin of man, the antistrophe adds, for it is by his vain hopes and foolish desires that he is led into trouble, according to the old saying that evil seems good to him whose mind is set on wickedness. Here we see that the feeling of dread of evil, which has been an undertone in previous choruses, a single thread interwoven with a different texture, comes to be the dominant tone; and so it remains, with but a single partial exception, through the rest of the play.

At the end of this choral song is an anapaestic stanza introducing Haemon. Kreon at once asks him on which side of the controversy he stands, to which Haemon gives an ingeniously ambiguous answer. Kreon construes it as positively in his favor, yet shows his inward doubt by going on to give his son a long lecture on his duty, proving by a variety of arguments the importance of pleasing one's father, and of maintaining the government under which one lives. Haemon then, in a speech of equal length, utters his views plainly, claiming for himself a right of independent judgment, telling his father how the citizens condemn the threatened punishment of Antigone, and urging him not to persist to the extreme of obstinacy in seeing only one side of the matter, and sticking to his own opinion. They go on from this, disputing in single verses, until both get thoroughly angry. Finally Kreon orders Antigone to be brought and put to death in presence of Haemon, upon which the latter rushes away, vowing never to see his father again. Kreon in his passion says that both the sisters shall die, but at the suggestion of the chorus admits that Ismene cannot be included in the penalty. But, instead of the death by stoning, which Antigone had heard was to be inflicted, he now substitutes death by starvation in an underground chamber, apparently as the more cruel form.

Here comes in a short choral song of two stanzas celebrating the resistless power of love, suggested apparently by the boldness which that passion had imparted to Haemon in standing up against his father's will. At its close an anapaestic stanza again announces the coming of Antigone on her way to death. It makes also the transition to the ensuing *kommos*, in which Antigone laments her fate, in lyric stanzas, and the chorus responds, comforting or rebuking her, first in anapaestic and then in iambic dimeters. It has seemed to some that these lamentations of Antigone's were tedious and protracted beyond the limits of good taste; to others, that they were out of character in the heroic girl who had dared to do the forbidden deed and then to defend it so bravely. The first of these criticisms I think has been sufficiently answered in the preface to President Woolsey's edition of the play. As to the other, it should be said that such laments seem natural to any human being in the immediate prospect of such a death, and that it would be unnatural for a young and tenderly reared woman to suppress them. Furthermore, there is nothing in them that implies the least repentance for her act. If Kreon had offered her pardon on condition of any form of recantation, we can have no doubt with what scorn she would have treated the offer. Her latest words show that she still thinks that what she did was right, and these laments are simply the natural utterances of grief at being cut off from life in all the freshness of her youth. It does not imply any failure of her courage, that she recognizes the horrors of the fate before her, and pours out her grief about it.

As might be expected, these laments are not very gratifying to Kreon's ear, and presently he comes out to stop them and hurry her on her way. But the poet allows her time for another long address in iambics. She greets the tomb which is to be her bridal chamber, and the members of her family who have died before her. She justifies her conduct in daring so much for her brother's sake. She appeals to the gods to convince her of error in what she has done, or to avenge the wrong she suffers. Finally Kreon threatens those in charge of her with punishment if they let her linger any more, and then at length she really goes. Her last words are, "See what I am suffering for having fulfilled a religious duty!" As she goes off, the chorus addresses to her the fifth *stasimon*. This belongs to the mythological type, so to call it, consisting wholly of an enumeration of mythical characters whose fate was in one point or another parallel to the one *à propos* of which they are mentioned. Here we have first Danaë, who was shut up in a tomb-like box, then Lykurgos, king of the Edones, who was imprisoned in a rock-cut chamber, and last, Kleopatra, wife of Phineus, and her two sons, who were likewise put in confinement, although she was of divine parentage.

At the end of this chorus, without the anapaestic announcement usual in this play, a new person appears. It is Teiresias, the blind seer, who comes unbidden to tell Kreon what his prophetic art has just been making known to him. By both methods of divination, the actions of birds and the condition of victims on the altar, he has learned that something is wrong, and he is convinced that the gods are offended by the fact that pieces of the unburied body of Polyneikes are brought near their altars by dogs and unclean birds. Therefore he advises Kreon, in much the same terms that Haemon had used, to lay aside his wrath and let the body be buried. But Kreon is in no mood for this. Forgetting how mistaken his former assumption had proved to be, that some one had bribed the guards to defeat his purpose, he at once makes the same assumption quite as confidently about the prophet, — that he has been bribed, — and declares violently that nothing shall make him swerve from his purpose, not even if the throne itself of Zeus be polluted by pieces of the corpse. They wrangle together for a few lines, and then Teiresias exercises the other function of his office, that of foretelling the future, and solemnly warns Kreon that within a short time he must give up a life out of his own family in exchange for the life of Antigone, and to atone for his offense against the powers of the world below in denying burial to the corpse. He then withdraws, leaving Kreon distressed and terrified by his prophecy. The more stubborn he has been, the more completely he now breaks down, as soon as he is really frightened. He turns to the chorus for advice, which it eagerly gives him, and in obedience to it he hurries away to release Antigone and to have the corpse duly buried.

While he is gone the chorus breaks out in a prayer to Bacchus to come and purify the city from its pollution. It is in the form of a *hyporchema*, or a song accompanied by a rapid dance movement. It refers to the titles of the god, enumerates the places which he most frequents and from some one of which they pray him to come, urges the claims of Thebes to his special favor, and closes with honorific descriptions of his glories. It is one of the best examples we have of such a combination of hymn and prayer, and gives a clear idea of the Greek mind in the attitude of devotion. Now the play hastens to its close. A messenger comes in and after some moralizing tells the fact of the death of Haemon. Haemon's mother Eurydike appears on her way to pray at the temple of Pallas, and overhearing the messenger's words requires of him a full account of what has happened. So he tells her how he went in attendance on her husband, and how they performed duly the funeral rites over the corpse. Thence they went to the prison of Antigone, but here they were too late. Antigone had hung herself, and Haemon was there mourning over her dead body. At sight of his father he drew his sword and rushed upon him, but when Kreon escaped by flight, he turned and threw himself upon his sword and so perished with his intended bride. At the end of his story, Eurydike, giving up her own useless visit to the temple, goes back into the palace without a word, in a way which seems ominous of evil.

Again, a detached anapaestic stanza announces a

THE *ANTIGONE* OF SOPHOKLES. 155

new comer, and Kreon enters, bearing the body of his son, and bitterly lamenting his death, which he confesses that he himself has caused. In the midst of his self-reproachings, a messenger comes out from the palace and informs him that his wife Eurydike has just committed suicide on hearing of the death of her son. This of course redoubles his grief, and the play closes, leaving him in this deserved misery, with a reflection by the chorus on the folly of such sinful and obstinate self-will.

JOTTINGS.

CHARACTER OF ANTIGONE.—Not an ideal woman, nor drawn directly from any Greek woman of the poet's time or in history; a figure of heroic stature, embodying and possessed by one principle or idea. Suppose a modern poet to try to give such a picture of Jael, or Judith; it would not be a pleasing picture. It is remarkable how the poet here seems to strive to soften by hints what in direct depicting he must make hard and severe; note her relation to Haemon, her apparent popularity as shown by what he says of public sentiment about her death, her occasional expressions of affection to her family, especially verses 897 ff. For it must be borne in mind that she is not an embodiment of sisterly love, though it is often said that she is. It is not primarily love to her brother that made her do her bold deed, but another sentiment, strange to us but very familiar and powerful in the Greek mind, that of the religious obligation of members of a family to the dead of the family. This is shown by her defense, vv. 45 ff. This fulfilment of duty naturally endears her to the dead members of the family, especially Polyneikes (vv. 81, 899 f.), and it also naturally implies love to them, but does not proceed wholly from that feeling. The common view that it does, belittles the heroic figure of Antigone by as much as a sentiment, even a natural and pure one, such as family affection, is in itself a less noble thing than a keen and strong sense of duty.

Antigone or Kreon right? A question much discussed at one time. Of possible combinations only three probable — Antigone all right and

Kreon all wrong; Antigone all right and Kreon partly so; each partly right and partly wrong. Last seems most probable. That a tragic conflict should interest us, it seems almost necessary that there should be some measure or, at least, appearance of right on each side. That a tragic hero should be the best possible, he or she ought to be a noble character with some fault or defect shown in play.

Is Antigone's deed a failure? see *Hellenica*.

The *Antigone* in many respects a typical Greek tragedy. — Almost no plot. (The *Oedipus Rex* a notable exception to this rule). — One leading character with no development; others as foils or opponents, — here three, as in *Prometheus, Oedipus Coloneus, Electra*. — Chorus in neutral position, advising moderation to both parties. Chorus of elders, as in *Persae, Agamemnon, Oedipus Rex, Oedipus Coloneus*. — In disposition of parts, *prologos, parodos, etc.*, quite regular. — Epic element in narrative.

Why is Ismene in *prologos?* No other fit confidant. — First coming of guard is natural in the story, — it serves to show Kreon's character in treatment of him. — The arrest of Ismene by bringing her again upon the stage enables the poet to show in a new light the character of Antigone. — Kreon sins against a law of family, and is punished in family. — Introduction of Haemon an invention of Sophokles.

Faults of play: Argumentation between Kreon and Haemon too much like wrangling in court. Something of it in all Sophokles's plays, but in *Philoktetes* it is not offensive. None of it in Aeschylus, unless in *Eumenides*, — but that is a court scene.

No motive for Kreon's going away at v. 326, or coming back at v. 386.

VII.

THE BEGINNING OF A WRITTEN LITERATURE IN GREECE.[1]

AN article on the above subject by Professor F. A. Paley in Fraser's Magazine for March, 1880, furnishes an occasion for some criticism and for a statement of the grounds of an opinion differing somewhat from the one there maintained. I will first state as briefly as possible the arguments and conclusions of Paley's article, with comments, and then present what evidence I can in favor of a different view.

Mr. Paley's general proposition is, that there is no evidence of the use of writing to multiply copies of books until a much later date than is ordinarily supposed. It is difficult to determine precisely to what date he would bring it down, for his statements do not agree with one another. In one place he speaks of "the times of the Alexandrine school of learning, when, *for the first time* (the italics are his), the use of papyrus and the practice of transcription became common." But a page or two later he says, "Books were no sooner introduced than they became both popular and cheap. Treatises on eloquence, as those

[1] Reprinted from Transactions of American Philological Association, 1880.

by Tisias and Corax, mentioned in the *Phaedrus*, the stories of Aesop, and the philosophical dogmas of Anaxagoras, could be bought at Athens, in the time of Plato, for a very small sum." It is not easy to see how books could be "popular and cheap in the time of Plato," a hundred years before the time when first "the use of papyrus and the practice of transcription became common." But we will take the alternative which involves least divergence from the common opinion, and suppose Mr. Paley to mean, as indeed the whole drift of the article indicates, that the use of writing for books did not become common in Greece until after 400 B.C., and in fact was hardly known at all before that date. I may say here at the outset that my own belief is, that it was introduced as much as fifty years earlier, and was fully established and familiar for some years before 400 B.C.

The first argument for Mr. Paley's view is drawn, he says, from "the singular, significant, and most important fact which, so far as I am aware, has never been noticed, that the Greek language, so copious, so expressive, not only has no proper verbs equivalent to the Roman *legere* and *scribere*, but has no terms at all for any one of the implements or materials so familiar to us in connection with writing (pen, ink, paper, book, library, copy, transcript, etc.), till a comparatively late period of the language." Then in a note he explains that "the Greek equivalent to *legere* means, to speak, and that to *scribere* means properly, to draw or paint." The latter "came to be used of writing because it (*i.e.*, writing) was at first an adjunct to descriptive painting." "The Greek had two verbs which indirectly express reading, but they are clumsy shifts, unworthy of so complete a language, the one meaning *recognoscere*, the other *sibi colligere*." I have quoted this in full because it seems so strange a process of reasoning that I could hardly trust myself to summarize it correctly. If it proves anything, it proves that the Romans began to read and write earlier, or at least earlier relatively to the development of their language, than the Greeks. No language, of course, can have a word for either of these ideas (or any other) before the thing expressed by the word is known to the speakers of the language, but it does not appear that the use of the compound form ($\dot{\epsilon}\pi\iota$-$\lambda\acute{\epsilon}\gamma o\mu\alpha\iota$) proves any less frequency or familiarity with the thing than the use of the simple form (*legere*). Further, *legere* has other senses besides *to read*, and apparently does not mean *to read* before the time of Cicero. On the other hand, as was suggested to me by Mr. F. B. Tarbell, $\lambda\acute{\epsilon}\gamma\omega$, at least once in Plato (*Theaet.* 143 C.), and repeatedly in the orators, has the sense *to read aloud*, to recite from a manuscript. No such inference as is drawn by Mr. Paley from the use of different stems or simple and compound forms in kindred languages has any validity. One might as well argue from the fact that the same stem in modern German means *to speak* (*reden*) and in modern English *to read*, that the Germans talked more than the English, and the English read more than the Germans. As to *scribere* and γράφειν, Mr. Paley arbitrarily assumes, without any reason, I think, that all the uses of γράφειν and its derivatives, before the Periklean age, refer to painting or to scratching on a hard surface. The truth is rather that γράφειν means both of these, and after writing with ink is introduced, means that too, and the special meaning in each case must be determined by other considerations. That *scribere* means only *to write*, indicates merely that the literature from which we learn its meaning belongs to a period when writing was a familiar art. The alleged absence of the words for pen, ink, paper, etc., will be referred to below.

How, then, it will be asked, is the existence of the earlier Greek literature, or rather the preservation of it to later times, to be explained? How is it that we have any fragments of the early historians, and the whole work of Herodotos and Thukydides? Mr. Paley anticipates this question, and answers that in his opinion, "authors of works laboriously wrote them on strips of wood, probably on a surface prepared with wax." These autograph copies were the only ones in existence, and the only way of publishing a book was by public readings from these copies. He doubts whether it would have been possible to procure for money a copy of the histories of Herodotos or Thukydides in the lifetime of the authors. His reason for this view is that he finds no proof that the earlier Greeks had any writing-material equivalent to our paper or parchment. There are, to be sure, several passages, to be cited presently, where the words for papyrus, paper, and parchment occur, but because they are brief passages, or the only instances, he seems to think they have no weight. Yet it would seem as if a single occurrence of the word *kerosene* in a book printed before 1846, or of *wigwam* in a book earlier than the discovery of America, would be enough to show knowledge of the existence of the thing denoted by the word.

Mr. Paley's next argument is the absence of reference in the writers of the Periklean age, particularly Herodotos, Thukydides, and Plato, to the works of their predecessors. Such reference, he thinks, would certainly have been made if the later writers had had access to copies of the earlier works, and the comparative absence of it proves that no such copies were within their reach.

There are, it is true, remarkably few references by name to previous writers in the early Greek literature, but Mr. Paley seems to have overlooked several passages in Herodotos, where it is clearly implied that he consulted some kind of records or accounts of the events he narrates, or descriptions of states whose form of government he speaks of. They are as follows: 6 : 55 καὶ ταῦτα μέν νυν περὶ τούτων εἰρήσθω· ὅτι δὲ ἐόντες Αἰγύπτιοι, καὶ ὅτι ἀποδεξάμενοι ἔλαβον τὰς Δωριέων βασιληίας, ἄλλοισι γὰρ περὶ αὐτῶν εἴρηται, ἐάσομεν αὐτά· τὰ δὲ ἄλλοι οὐ κατελάβοντο, τούτων μνήμην ποιήσομαι, and then he goes on to speak of the privileges and functions of the Spartan kings.

9 : 81 ὅσα μὲν νῦν ἐξαίρετα τοῖσι ἀριστεύσασι αὐτῶν ἐν Πλαταιῇσι ἐδόθη, οὐ λέγεται πρὸς οὐδαμῶν, δοκέω δ' ἔγωγε καὶ τούτοισι δοθῆναι. A similar expression occurs in 8 : 133 ὅ τι μὲν βουλόμενος ... ταῦτα ἐνετέλλετο, οὐκ ἔχω φράσαι· οὐ γὰρ λέγεται· δοκέω δ' ἔγωγε κτλ. These passages plainly indicate that he had access, not merely to inscriptions and formal public records, but to writings prepared for the information of inquirers, and discussing the motives of actions as well as describing the early history of states. (The use of authorities by Herodotos is treated by Rawlinson in his Introduction, chapter II.) But it remains true, as Mr. Paley says, that there are exceedingly few quotations by name of these earlier writers.

Plato quotes Akusilaos once, Thukydides quotes Hellanikos once, Herodotos refers to Hekataeos three or four times — but beyond these few instances there is no recognition by these writers of the many persons who are said to have written prose before their time. Here Mr. Paley touches upon a singular fact which certainly is not easy of explanation. The most striking instance of it, perhaps, is the case of Thukydides, who is not mentioned, I believe, by any writer whose works we have, earlier than Dionysios of Halikarnassos, in the last century before the Christian era. But this fact will not bear the interpretation Mr. Paley puts upon it. It is true also in the next century, when books were common. Aristotle does not mention Hekataeos, Hellanikos, Akusilaos, Thukydides, or Xenophon. Plato does not quote from Xenophon, nor Xenophon from Plato.[1] A similar failure appears in the argument which Mr. Paley bases upon the statement in the *Phaedros* of Plato, that Lysias was taunted with being a λογογράφος, *speech-writer*, as almost the same with being a sophist. Mr. Paley regards this as "satirizing a practice which was then beginning to come into vogue." But the same contempt for λογογράφοι and σοφισταί together is expressed in Dem. *de Falsa Legatione*, a speech delivered in 342 B.C., long after the use of writing must have been familiar. It is plain that it is not the mere *writing* of the speech that is objected to, but the professional composition of speeches for others to use.

As the lack of reference to previous writers is mere negative evidence, Mr. Paley supplements it by the fact that Thukydides, in attempting to sketch the early history of Greece, is obliged to rest upon "inference, memory, hearsay." He has no current written literature to appeal to, and this is made to show that the previous historians, Herodotos and his predecessors, were not accessible to him. Indeed, Mr. Paley distinctly says, "Thukydides does not seem to have known Herodotos at all." These statements, which will surprise every Greek scholar, are founded on passages in the first book, sections 1, 9, 20, 21.

[1] Westermann (on Dem. *Ol.* 3: 21) remarks upon the habit of the orators of referring for matters of history to tradition rather than to written records, and explains it as due to a desire to identify themselves as much as possible with the average hearer, assuming no more knowledge than he would have.

They ignore the language of that "single reference" to Hellanikos in 1 : 97, which Mr. Paley repeatedly mentions but nowhere quotes. It deserves to be quoted in full from its clear evidence on this point. ἔγραψα δὲ αὐτὰ (*i.e.*, the outline of the growth of the Athenian empire after the Persian war) ... διὰ τόδε, ὅτι τοῖς πρὸ ἐμοῦ ἅπασιν ἐκλιπὲς τοῦτο ἦν τὸ χωρίον καὶ ἢ τὰ πρὸ τῶν Μηδικῶν Ἑλληνικὰ ξυνετίθεσαν ἢ αὐτὰ τὰ Μηδικά· τούτων δὲ ὅσπερ καὶ ἥψατο ἐν τῇ Ἀττικῇ ξυγγραφῇ Ἑλλάνικος, βραχέως τε καὶ τοῖς χρόνοις οὐκ ἀκριβῶς ἐπεμνήσθη. "I have written this outline for this reason, because all my predecessors have neglected this period and composed either a history of Greece before the Persian wars, or of the Persian wars themselves; and the one who did touch on this period in his history of Attika, Hellanikos, made but a brief record without strict chronological accuracy." It is clear from this, (1) that he knew the works of several predecessors in full, so that he could tell what periods they treated and in what way; (2) that he knew Herodotos's work, for no one else so far as we know, wrote so full a history of the Persian wars; and (3) that he expected readers to look in *their* histories for information on that period, and, failing to find it, to have recourse to his. (Cf. 1 : 23 διότι δ' ἔλυσαν, τὰς αἰτίας προέγραψα πρῶτον καὶ τὰς διαφοράς, τοῦ μή τινα ζητῆσαί ποτε ἐξ ὅτου τοσοῦτος πόλεμος τοῖς Ἕλλησι κατέστη.) How, then, are those other passages to be understood, wherein he speaks as if obliged to rest on tradition and without any previous authorities to refer to? Simply by recognizing the evident fact that he did not regard his predecessors as authorities. He had formed for himself a new standard of historic evidence — and, tested by that standard, the works of his predecessors could not command his confidence. He refused to trust such material as Herodotos used, and he means by this language to indicate that in his view all previous so-called histories rested merely on tradition. It can hardly be doubted that he included Herodotos, as well as Hellanikos and Hekataeos among the λογογράφοι, "who composed rather to please the ear than with a view to truth."

One other point in Mr. Paley's article deserves notice. He supposes that the stories, histories, and philosophic teachings of the early Greeks were a purely oral literature, and that they were put into writing eventually from the dictation of the pupils and followers of their authors — and that thus it happens that the *writings* of the early philosophers and historians are referred to. It would seem from this suggestion that Mr. Paley can hardly have ever looked into the fragments of the early historians. He would have found a reasonably large number of such fragments, from Hekataeos, Charon, Xanthos, Hellanikos, and Akusilaos, preserving in many cases apparently the original words of the authors, and quoted from works of some extent, of which the titles are given. He would have seen also that the matter of these quotations and the style are such as to make it impossible to imagine them orally delivered and preserved by memory until after the lapse of years writing was introduced. It is, I think, really impossible to suppose that such matter as makes up the "Europe" and "Asia" of Hekataeos, for example, can ever have been delivered orally by a master to a group of listening pupils. For it consists largely, if we may judge by the fragments preserved, of a list of names of towns — hardly more than the simple name in many cases, with a brief indication of the locality. One example, taken almost at random, may show the character of a multitude: Steph. Byz. Χάλαιον· πόλις Λοκρῶν· Ἑκαταῖος Εὐρώπῃ· "μέτα δὲ Λοκροί, ἐν δὲ Χάλαιον πόλις, ἐν δὲ Οἰανθὴ πόλις." (Müller, F. H. G., 83.) One might as well commit the dictionary to memory as matter like this, without help of metre or of connection. Not only could it not be committed to memory, but we may rightly argue from the subject matter that it would not be composed before the time when the idea of a book had become a familiar idea. The making of such a record does not belong to the age of epic narration, nor to that of lyric song, nor to that of oral speculative discourse, but to that in which history begins — when men first recognize the value of facts preserved in writing and begin to regard matter as well as form. That gave rise to a prose style, and thus also made writing necessary. What could induce a man to put together such a string of bare facts as this, except the desire to preserve the knowledge for the information of others in such a form that they could consult it? We cannot imagine Hekataeos as delivering orally such matter as this to a company of hearers. We must suppose that it was written out from the first, and either kept by him for consultation, or, as seems more likely, copied out as a whole or in part for the convenience of those whose interests, of trade or colonization, made them willing to pay for the work.

I come now, omitting several minor points in Mr. Paley's article which are open to criticism, to the evidence upon which I rely to carry back the extensive use of writing to the middle of the fifth century before Christ. It may seem the more worth while to do this because, so far as I can ascertain, this precise point has not been fully illustrated in any easily accessible work. Several of the passages cited are referred to in Mr. Paley's article, but have in his view little or no importance. The passages are arranged as nearly as possible in chronological order.

Pind. *Ol.* XI. 1 ff.

> τὰν Ὀλυμπιονίκαν ἀνάγνωτέ μοι
> Ἀρχεστράτου παῖδα πόθι φρενὸς
> ἐμᾶς γέγραπται.

This appears to be, as Mr. Paley says, the earliest instance of ἀναγιγνώσκω meaning *to read*. It is more than a mere instance of the word, for it shows it in connection with γράφειν meaning *to write* or *engrave*, and both together in a metaphor, which would hardly be natural or intelligible, unless the two ideas in this association were so familiar as to be caught at once by hearers of the ode. The practice of reading written words must have been not the secret art of a few, but in some degree a part of common life, before a poet could thus casually refer to it. Unfortunately, this ode cannot be precisely dated, though it must belong some years before 440 B.C., near which time the poet died. The same metaphor occurs repeatedly in Aeschylos (*e.g.*, *Prom.* 989, *Supp.* 991, *Cho.* 441).

Aesch. *Supp.* 946f.

> ταῦτ᾽ οὐ πίναξίν ἐστιν ἐγγεγραμμένα
> οὐδ᾽ ἐν πτυχαῖς βίβλων κατεσφραγισμένα.

The second of these lines Mr. Paley brackets in his third edition, on the ground of the metre, though the fault had not attracted his notice before. No other editor has ever suspected its genuineness, and many other lines no less open to objection stand unchallenged (*e.g.*, *Supp.* 465, 931, 1016). It can hardly be doubted, I think, that the desire to get rid of the evidence of the line on the question of the use of writing sharpened Mr. Paley's sense of its faulty metre. For it plainly testifies to the familiar use of papyrus, folded and sealed, at the same time with that of wax-covered tablets. The date of the *Supplices* is not known, but from its structure it seems to be one of the earlier plays of Aeschylos, and no

BEGINNING OF A WRITTEN LITERATURE. 169

one, so far as I know, has placed it later than 460 B.C.

The next witness is Herodotos, whose history is supposed, from the latest incident referred to in it, to have been finished in its present form by about the year 425. Of course the material for it was gathered in great measure before this date, and his numerous references (1 : 123, 125 ; 3 : 42, 123, 128) to writing upon papyrus, γράφειν ἐς βιβλίον, though they may all refer to short memoranda or notes, yet imply familiar and frequent use of writing before his time. But the particular passage which I quote indicates much more than that. He says, in 5:58: καὶ τὰς βίβλους διφθέρας καλέουσι ἀπὸ τοῦ παλαιοῦ οἱ Ἴωνες, ὅτι κοτὲ ἐν σπάνι βίβλων ἐχρέοντο διφθέρῃσι αἰγέῃσί τε καὶ οἰέῃσι· ἔτι δὲ καὶ τὸ κατ' ἐμὲ πολλοὶ τῶν βαρβάρων ἐς τοιαύτας διφθέρας γράφουσι. "And the Ionians from old usage give the name διφθέραι (skins) to sheets of papyrus, because when papyrus was scarce they used to use instead skins of goat and sheep; and still even in my day many uncivilized peoples use such skins for writing." This passage proves that papyrus was the usual material for writing, as much so as paper in our day, and that it had been so for a long time. Also, that it was ordinarily plentiful among the Ionians of Asia Minor and the Greeks generally in the time of Herodotos. He explains the local use of the word διφθέραι (skins) as a name for papyrus, as arising from a local scarcity of papyrus. Whether the explanation is correct or not, it plainly shows that the writer thought of papyrus as the common thing for everybody to write on—at least among civilized Greeks, for he adds that some uncivilized peoples still used skins or parchment. In my view this passage alone supplies fully that which Mr. Paley desiderates, viz., some mention of the use of papyrus as a writing material. It fully supports the statements of Grote and Hayman, which Mr. Paley characterizes as "unsupported by evidence."

In connection with this passage should be mentioned the occurrence in certain comic poets, of about the same time with Herodotos, of words implying the commonness in ordinary life of writing and apparently of books. These words are mentioned by Pollux (vii. 210). Thus he ascribes to Kratinos, who died about 422 B.C., the word βιβλιογράφος, and quotes (ix. 47) from Eupolis, whose latest known play was given in 412 B.C., the phrase οὗ τὰ βιβλία ὤνια, "where is the book-market." Other similar words occur in later poets. In Aristophanes there are repeated references to books. Thus in the *Frogs* (405 B.C.), verse 943,

(ἴσχνανα τὴν τραγῳδίαν)
χυλὸν διδοὺς στωμυλμάτων ἀπὸ βιβλίων ἀπηθῶν—

"I reduced tragedy in flesh by feeding her on a porridge of moral maxims drawn from books." And again, *Frogs* 1113 ff., where the chorus addresses the two poets just as they are going to compare their poetic styles:

BEGINNING OF A WRITTEN LITERATURE. 171

ἐστρατευμένοι γάρ εἰσι,
βιβλίον τ' ἔχων ἕκαστος μανθάνει τὰ δέξια —

"(Fear not that the audience will not understand your jokes,) for they have been disciplined and every man has his book too and learns wisdom out of it." These are all instances of reference to books in general, but we have one from the same time which names a particular book. It is the passage already quoted from Thukydides (1 : 97). I may repeat here the translation of it : "I have written this sketch for this reason, viz., because all my predecessors have neglected this period and composed either a history of Greece before the Persian wars, or of those wars themselves ; and the one who did touch on this period in his history of Attika, Hellanikos, made but a brief record without strict chronological accuracy." Here we have reference to several histories, with implied knowledge of their contents, and special reference to one of which the title is given ἡ Ἀττικὴ ξυγγραφή, being, I take it, a mere paraphrase for ἡ Ἀτθίς, under which name the book is quoted by later writers. This passage must have been written before 400 B.C., and probably was written as early as between the Peace of Nikias (422 B.C.) and the Sicilian expedition (415 B.C.). It supplies, from an almost contemporary source, clear proof of the early existence of written copies of the first Greek attempts at history, the existence of which has already been inferred from the subject matter and style of the histories as seen in the abundant fragments of them.

Another passage of Aristophanes, as commonly interpreted, mentions by title a copy of a particular book. It is in the *Frogs*, 52 ff. :

καὶ δῆτ' ἐπὶ τῆς νεὼς ἀναγιγνώσκοντί μοι
τὴν Ἀνδρομέδαν πρὸς ἐμαυτὸν ἐξαίφνης πόθος
τὴν καρδίαν ἐπάταξε.

Mr. Paley does not overlook this passage, but evades the force of it against his theory by explaining it as referring to the name of a ship. In his view, Dionysos sitting on his own ship saw another near by with the name "Andromeda" painted on its stern or bow, and, as his eye rested on that name and he idly read it over and over, it reminded him of the play of Euripides bearing the same name and so called up in him a longing for the poet. It is not possible, perhaps, to show that this explanation is certainly and necessarily a mistaken one, yet surely the common explanation, that he was reading a copy of the play, is more natural and probable. The tense of ἀναγιγνώσκοντι and the addition of πρὸς ἐμαυτόν to it, are indications in favor of this view. The passage so understood shows that it was nothing strange in 405 B.C. for a man going to serve in the Athenian fleet to take with him a copy of some favorite author or book.

As to the material on which such books were written, we have, besides the passage from Herodotos already quoted, a line from Plato Comicus, quoted by Pollux (vii. 210), which proves the use of the later word for paper in his time (425–395 B.C.) :

τὰ γραμματεῖα τούς τε χάρτας ἐκφέρων,

"bringing out the tablets and the sheets of paper." With this should be put the passage from the other and greater Plato (*Phaedros*, 276 C.), where he says : οὐκ ἄρα σπουδῇ αὐτὰ ἐν ὕδατι γράψει μέλανι σπείρων διὰ καλάμου — "he will not then laboriously write them in water, sowing (his seed of truth) with ink through a pen." The date of the *Phaedros* cannot be certainly determined, though some scholars have maintained that it must have been one of Plato's earliest writings. In any case we have here, not far from 400 B.C. on either side, mention of pen, ink, and paper (made, of course, from papyrus), and I would call attention to the perfectly incidental, matter-of-course character of the reference to pen and ink, in an illustration, in this last passage. It is not so that a writer would speak of a new instrument, just introduced and known to few persons.

The passages so far cited, except the last, have been all taken from writers or writings prior to 400 B.C. It seems proper, however, to add some from Xenophon and Plato, whose writings probably all belong after that date. It will be seen that one of these certainly and others probably involve recognition of books as easily accessible before that date. The lives of these two men extend from about 430

B.C. to about 355 B.C., but their writings were probably all composed after 400 B.C. It is a great misfortune, especially in the case of Plato and with regard to the history of his philosophical opinions, that the chronological order of these works cannot be ascertained. But I think it is fair to accept his incidental references to the existence and use of books as evidence of the facts within the first twenty-five years after 400 B.C.

I begin with the passages from Xenophon:

Mem. I. 6. 14 καὶ τοὺς θησαυροὺς τῶν πάλαι σοφῶν ἀνδρῶν οὓς ἐκεῖνοι κατέλιπον ἐν βιβλίοις γράψαντες, ἀνελίττων κοινῇ σὺν τοῖς φίλοις διέρχομαι. "And the treasures of the wise men of old which they have left behind them in written books, I open and read over in company with my friends." It is Sokrates who speaks here, and the conversation in which the words occur, Xenophon explicitly tells us that he himself heard. It must have occurred then before his departure from Athens to join Kyros on his illfated expedition, that is, before 401 B.C. If there is any historic truth in the *Memorabilia*, it would be in a passage thus commended to us by the author himself, and I hardly see how we could ask for clearer or better evidence that books were easily obtained in the lifetime of Sokrates. That they were to be obtained for money appears from another passage:

Xen. *Mem.* IV. 2. 1 (ὁ Σωκράτης κατέμαθεν) Εὐθύδημον τὸν καλὸν γράμματα πολλὰ συνειλεγμένον ποιητῶν τε καὶ σοφιστῶν τῶν εὐδοκιμωτάτων. ... 8. εἰπέ μοι, ἔφη, ὦ Εὐθύδημε, τῷ ὄντι, ὥσπερ ἐγὼ ἀκούω, πολλὰ γράμματα συνῆχας τῶν λεγομένων σοφῶν ἀνδρῶν γεγονέναι; Νὴ τὸν Δία, ἔφη, ὦ Σώκρατες· καὶ ἔτι γε συνάγω, ἕως ἂν κτήσωμαι ὡς ἂν δύνωμαι πλεῖστα. Νὴ τὴν "Ηραν, ἔφη ὁ Σωκράτης, ἄγαμαί γέ σου, διότι οὐκ ἀργυρίου καὶ χρυσίου προείλου θησαυροὺς κεκτῆσθαι μᾶλλον ἢ σοφίας... 10. Τί δὲ δὴ βουλόμενος ἀγαθὸς γενέσθαι, ἔφη, ὦ Εὐθύδημε, συλλέγεις τὰ γράμματα; ἐπεὶ δὲ διεσιώπησεν ὁ Εὐθύδημος, σκοπῶν ὅτι ἀποκρίναιτο, πάλιν ὁ Σωκράτης, Ἆρα μὴ ἰατρός; ἔφη· πολλὰ γὰρ καὶ ἰατρῶν ἐστι συγγράμματα. (Sokrates learned) "that Euthydemos, a noble youth, had collected many writings of the most eminent poets and learned men. ... 'Tell me, Euthydemos,' said he, 'have you really, as I am told, collected many writings of those who have been eminent for wisdom?' 'Certainly, Sokrates,' said he, 'and I am still collecting in order to get as many as I possibly can.' 'By Hera,' said Sokrates, 'I am delighted with you, because you have not preferred the possession of treasures of money to that of treasures of wisdom.... But what is it that you want to excel in, Euthydemos,' said he, 'that you are collecting books?' And when Euthydemos was silent, considering what answer to make, 'Is it in medicine?' asked Sokrates, 'for there are many books on that subject.'" Here the praise given to the preference of wisdom over wealth shows that the books had been obtained by purchase. Though this conversation is not vouched for, as the other is, by Xenophon's statement that he heard it, yet it probably has historic reality, and if so, must have occurred before 400 B.C., and probably some years before the time of the Thirty (404 B.C.).

Another passage shows that books were exported to the Greek colonies on the Euxine Sea:

Xen. *Anab.* VII. 5. 14 (The Ten Thousand on their way home come to Salmydessos and find there many spoils of wrecks on that dangerous coast.) ἐνταῦθα εὑρίσκονται πολλαὶ μὲν κλῖναι, πολλὰ δὲ κιβώτια, πολλαὶ δὲ βίβλοι γεγραμμέναι, καὶ τἄλλα πολλὰ ὅσα ἐν ξυλίνοις τεύχεσι ναύκληροι ἄγουσιν. "There were found many bedsteads, and many chests, and many written books, and quantities of other things of all kinds that shipmasters convey in wooden cases." The word γεγραμμέναι here is wanting in some inferior manuscripts, but all the later editors (L. Dindorf, Krüger, Rehdantz, Vollbrecht, Sauppe) take it into their text without question. These works of Xenophon were probably written after 390 B.C., but the evidence in these quoted passages all refers to facts occurring before 400 B.C. Of these passages Mr. Paley takes no notice whatever.

I add now a few passages from Plato, not as proof of the existence of written books before 400 B.C., — for the writings of Plato are of too uncertain date and presumably too late for that, — but as indicating how common and accessible books were, and on how great a variety of subjects they were composed, within the first thirty or forty years after that date. It may be legitimate to reason backwards from this fact and infer something like a similar rapidity in the spread of the new practice before 400 B.C., and thus get a confirmation of what we might conclude from the passages already quoted.

Apol. 26 D Ἀναξαγόρου οἴει κατηγορεῖν, ὦ φίλε Μέλητε, καὶ οὕτω καταφρονεῖς τῶνδε καὶ οἴει αὐτοὺς ἀπείρους γραμμάτων εἶναι, ὥστε οὐκ εἰδέναι ὅτι τὰ Ἀναξαγόρου βιβλία τοῦ Κλαζομενίου γέμει τούτων τῶν λόγων; Here it will be observed that Plato represents Sokrates as saying that it would impute illiteracy or at least strange want of knowledge of current literature to the jurors, men chosen by lot, some five hundred perhaps in number, from all ranks of the citizens, to suppose them ignorant of the fact that "the books of Anaxagoras teem with such doctrines" as the accuser charged him with holding. "The books of Anaxagoras," one would think, must have been easily within the reach of the people when this could be said. The next succeeding sentence, in which reference is made to "buying from the orchestra, for a drachma at the highest, power to ridicule Sokrates if he claims these doctrines as original with him," is so much disputed as to its precise meaning that it is better not to use it in evidence here.

Phaed. 97 C ἀλλ' ἀκούσας μέν ποτε ἐκ βιβλίου τινός, ὡς ἔφη, Ἀναξαγόρου ἀναγιγνώσκοντος κτλ.

98 B καὶ οὐκ ἂν ἀπεδόμην πολλοῦ τὰς ἐλπίδας, ἀλλὰ πάνυ σπουδῇ λαβὼν τὰς βίβλους ὡς τάχιστα οἷός τ' ἦν ἀνεγίγνωσκον.

Sympos. 177 B ἔγωγε ἤδη τινὶ ἐνέτυχον βιβλίῳ, ἐν ᾧ ἐνῆσαν ἅλες ἔπαινον θαυμάσιον ἔχοντες πρὸς ὠφέλειαν, καὶ ἄλλα τοιαῦτα συχνὰ ἴδοις ἂν ἐγκεκωμιασμένα.

Gorg. 462 B Πῶλος. Ἀλλὰ τί σοι δοκεῖ ἡ ῥητορικὴ εἶναι; Σωκρ. Πρᾶγμα ὃ φῂς σὺ ποιῆσαι τέχνην ἐν τῷ συγγράμματι ὃ ἐγὼ ἔναγχος ἀνέγνων.

518 B Μίθαικος ὁ τὴν ὀψοποιίαν συγγεγραφὼς τὴν Σικελικήν. (Mithaikos, author of the "Handbook of Sicilian Cookery.")

Protag. 325 E. οἱ δὲ διδάσκαλοι τούτων τε ἐπιμελοῦνται, καὶ ἐπειδὰν αὖ γράμματα μάθωσι καὶ μέλλωσι συνήσειν τὰ γεγραμμένα, . . παρατιθέασιν αὐτοῖς ἐπὶ τῶν βάθρων ἀναγιγνώσκειν ποιητῶν ἀγαθῶν ποιήματα καὶ ἐκμανθάνειν ἀναγκάζουσιν. (If the boys had copies of Homer and Hesiod to learn lessons from in school, one would suppose their fathers might have had them to read.)

Phaedr. 228 D Σωκρ. Δείξας γε πρῶτον, ὦ φιλότης, τί ἄρα ἐν τῇ ἀριστερᾷ ἔχεις ὑπὸ τῷ ἱματίῳ. τοπάζω γάρ σε ἔχειν τὸν λόγον αὐτόν. (And so he had a copy of Lysias' speech, which he presently reads.)

230 D. . . σὺ ἐμοὶ λόγους οὕτω προτείνων ἐν βιβλίοις τήν τε Ἀττικὴν φαίνει περιάξειν ἅπασαν καὶ ὅποι ἂν ἄλλοσε βούλῃ.

273 A τόν γε Τισίαν αὐτὸν πεπάτηκας ἀκριβῶς. (This same phrase, πεπατηκέναι τινά, *to be familiar with an author*, occurs in the *Birds* of Aristophanes (v. 471) οὐδ' Αἴσωπον πεπάτηκας. It seems to imply almost necessarily the use of a copy of the author's

works. The *Birds* came out in 415 B.C. Mr. Paley speaks of this phrase as new in the time of Plato's literary activity.)

276 C. (The passage speaking of pen and ink, already quoted.)

Theaet. 152 A Σωκρ. φησὶ γάρ που πάντων χρημάτων μέτρον ἄνθρωπον εἶναι... ἀνέγνωκας γάρ που; Θεαίτ. Ἀνέγνωκα καὶ πολλάκις.

162 A εἰ ἀληθὴς ἡ ἀλήθεια Πρωταγόρου, ἀλλὰ μὴ παίζουσα ἐκ τοῦ ἀδύτου τῆς βίβλου ἐφθέγξατο.

166 C οὐ μόνον αὐτὸς ὑηνεῖς, ἀλλὰ καὶ τοὺς ἀκούοντας τοῦτο δρᾶν εἰς τὰ συγγράμματά μου ἀναπείθεις.

Soph. 232 D Ξέν. Τά γε μὴν περὶ πασῶν τε καὶ κατὰ μίαν ἑκάστην τέχνην, ἃ δεῖ πρὸς ἕκαστον αὐτὸν τὸν δημιουργὸν ἀντειπεῖν, δεδημοσιωμένα που καταβέβληται γεγραμμένα τῷ βουλομένῳ μαθεῖν. Θεαίτ. Τὰ Πρωταγόρειά μοι φαίνει περί τε πάλης καὶ τῶν ἄλλων τεχνῶν εἰρηκέναι. Ξέν. Καὶ πολλῶν γε, ὦ μακάριε, ἑτέρων.

Polit. 293 A τοὺς ἰατροὺς δὲ οὐχ ἥκιστα νενομίκαμεν, ἐάν τε ἑκόντας ἐάν τε ἄκοντας ἡμᾶς ἰῶνται, .. καὶ ἐὰν κατὰ γράμματα ἢ χωρὶς γραμμάτων, .. πάντως οὐδὲν ἧττον ἰατρούς φαμεν κτλ.

Parmen. 128 D διὰ τοιαύτην δὴ φιλονεικίαν ὑπὸ νέου ὄντος ἐμοῦ ἐγράφη, καί τις αὐτὸ ἔκλεψε γραφέν, ὥστε οὐδὲ βουλεύσασθαι ἐξεγένετο, εἴτ' ἐξοιστέον αὐτὸ εἰς τὸ φῶς εἴτε μή.

In these passages we see that books were so common in Plato's time that not to know the contents of a certain one would prove a man deficient in education, — that they were put before schoolboys to learn lessons out of, — that particular ones were read again and again by the same person, — that there were books on rhetoric, on the uses of salt, on cookery, on medicine, on wrestling, and, in a word, on all arts, — that once a book was stolen and circulated while the author was still deliberating about publishing it, — that a man overheard another reading from a book and immediately got hold of the book to read it for himself. If now the use of books was so general in all circles of life in Plato's time, the first thirty or forty years after 400 B.C., and if, as we have previously seen, mention of reading and writing, of tablets, papyrus, and parchments goes back to about 450 B.C., and the mention of books and of book-writers (copyists) and book-selling comes along between 420 and 405 B.C., can it be supposed that so quick-witted a people as the Athenians, so interested especially in every stimulus to mental activity, failed to see the capabilities of this contrivance and to make use of it in that earlier period?

I may be permitted in conclusion briefly to restate the evidence as to that earlier period. We have in Pindar before 450 B.C., a metaphor drawn from the arts of writing and reading. We have in Aeschylos, before 460 B.C., repeatedly the metaphor from writing, and once a mention of tablets and of papyrus. We have in Herodotos, before 425 B.C., frequent reference to writing on papyrus, and once a recognition of that as the usual material for writing, occasionally supplemented by parchment. We have abundant fragments of Hekataeos (540–480 B.C.) and other early historians, in a style of composition that forbids the idea of oral transmission. We have from the comic poets Kratinos (before 420 B.C.), Eupolis (before 412 B.C.), and Plato (probably before 405 B.C.), fragments containing mention of book-writing, paper, and book-selling. We have from Aristophanes (in plays down to 405 B.C.) reference to books as used by authors and readers, and consulted by his own audience. We have in Thukydides (probably before 405 B.C.) reference to the works of his predecessors implying knowledge of their contents on his part, and a suggestion that other historical inquirers would consult his own work as he had theirs. Finally we have in Xenophon (in reference to a time before 400 B.C.) mention of books as read among a company of friends, as bought by a collector of a library, and as exported to the shores of the Euxine sea. Now in view of this evidence, recognizing the fragmentary character of the remains we have of the literature of the fifth century before Christ, are we not justified in holding that the use of writing on papyrus for the purpose of preserving and multiplying copies of works of literature began as early as the middle of that century and rapidly grew to be a familiar matter of common life before its end?

It will be observed that I have confined myself to the production of the evidence attainable on my subject with only the necessary explanation of it. My purpose has been simply to bring together all the passages which I could find containing real evidence, in the hope that the collection, not elsewhere made so far as I know, might be of service to any one who wishes to ascertain the facts.

Latin Text-Books.

		INTROD. PRICE
ALLEN & GREENOUGH:	Latin Grammar	$1.12
	Latin Composition	1.12
	Caesar (four books, with vocabulary)	1.12
	Sallust's Catiline	.60
	Cicero, 13 orations (or 8 orations with vocabulary)	1.12
	Cicero de Senectute	.50
	Ovid (with vocabulary)	1.40
	Virgil (Bucolics and 6 Books of the Æneid)	1.12
	Preparatory Course of Latin Prose	1.40
ALLEN	Latin Primer	.90
	New Latin Method	.90
	Introduction to Latin Composition	.90
	Latin Reader	1.40
	Latin Lexicon	.90
	Remnants of Early Latin	.75
	Germania and Agricola of Tacitus	1.00
BLACKBURN	Essentials of Latin Grammar	.70
	Latin Exercises	.60
	Latin Grammar and Exercises (in one volume)	1.00
CROWELL	Selections from the Latin Poets	1.40
CROWELL & RICHARDSON:	Brief History of Roman Lit. (BENDER)	1.00
GREENOUGH	Virgil: —	
	Bucolics and 6 Books of Æneid (with Vocab.)	1.60
	Bucolics and 6 Books of Æneid (without Vocab.)	1.12
	Last 6 Books of Æneid, and Georgics (with notes)	1.12
	Bucolics, Æneid, & Georgics (complete, with notes)	1.60
	Text of Virgil (complete)	.75
	Vocabulary to the whole of Virgil	1.00
GINN & HEATH:	Classical Atlas and Geography (cloth)	2.00
HALSEY	Etymology of Latin and Greek	1.12
	Classical Wall Maps (three or more), each	3.50
KEEP	Essential Uses of the Moods in Greek and Latin	.25
KING	Latin Pronunciation	.25
LEIGHTON	Latin Lessons	1.12
MADVIG	Latin Grammar (by Thacher)	2.25
PARKHURST	Latin Verb	.35
PARKER & PREBLE:	Handbook of Latin Writing	.50
SHUMWAY	Latin Synonymes	.30
STICKNEY	Cicero de Natura Deorum	1.40
TETLOW	Inductive Latin Lessons	1.12
TOMLINSON	Manual for the Study of Latin Grammar	.20
WHITE (J. W.)	Schmidt's Rhythmic and Metric	2.50
WHITE (J. T.)	Junior Students' Latin-English Lexicon (mor.)	1.75
	English-Latin Lexicon (sheep)	1.50
	Latin-English and English-Latin Lexicon (sheep)	3.00
WHITON	Auxilia Vergiliana; or, First Steps in Latin Prosody	.15
	Six Weeks' Preparation for Reading Cæsar	.35

Copies sent to Teachers for Examination, with a view to Introduction, on receipt of Introduction Price.

Send for description of our new Illustrated Caesar (seven books).

GINN & COMPANY, Publishers,
BOSTON, NEW YORK, AND CHICAGO.

GREEK BOOKS.[2]

Allen	Medea of Euripides	$1.00
College Series of Greek Authors.... See D'Ooge, Dyer, Humphreys.		
D'Ooge	Sophocles' Antigone: *Text and Notes*	.95
	Text only	.45
Dyer	Plato's Apology and Crito: *Text and Notes*	.95
	Text only	.45
Flagg	Hellenic Orations of Demosthenes	1.00
	Anacreontics	.35
	Seven against Thebes	1.00
Goodwin	Greek Grammar	1.50
	Greek Reader	1.50
	Greek Moods and Tenses	1.50
	Selections from Xenophon and Herodotus	1.50
Goodwin & White:	Anabasis	1.00
	Anabasis (*with Vocabulary*)	1.50
Humphreys	Aristophanes' Clouds: *Text and Notes* ...	.95
	Text only	.45
Keep	Essential Uses of the Moods	.25
Kendrick	Greek at Sight	.15
Leighton	New Greek Lessons	1.20
Liddell & Scott..	Abridged Greek-English Lexicon	1.90
	Unabridged Greek-English Lexicon	9.40
Seymour	Selected Odes of Pindar	1.40
Sidgwick	Greek Prose Composition	1.50
Tarbell	Philippics of Demosthenes	1.00
Tyler	Selections from Greek Lyric Poets	1.00
White	First Lessons in Greek	1.20
	Schmidt's Rhythmic and Metric of the Classical Languages	2.50
	Œdipus Tyrannus of Sophocles	1.12
	Stein's Dialect of Herodotus	.10
Whiton	Orations of Lysias	1.00

Copies sent to teachers for examination, with a view to Introduction, on receipt of Introduction Price given above.

GINN & COMPANY, Publishers.

BOSTON.   NEW YORK.   CHICAGO.

www.ingramcontent.com/pod-product-compliance
Lightning Source LLC
Chambersburg PA
CBHW032142160426
43197CB00008B/745